The New Economy of Oil

Research by the Energy and Environment Programme of the Royal Institute of International Affairs is supported by generous contributions of finance and technical advice from the following organizations:

Amerada Hess Ltd
Anglo American plc
BG Group
Blue Circle Industries plc
BP Amoco plc
British Energy plc
British Nuclear Fuels plc
Department of Environment, Transport & the Regions (UK)
Department of Trade & Industry (UK)
ExxonMobil
Foreign & Commonwealth Office (UK)
Lasmo plc
National Grid plc
Osaka Gas Co. Ltd
Powergen plc
Shell UK
Statoil
Tokyo Electric Power Co. Inc.
TotalFinaElf
TXU Europe Group plc
The Wrigley Company

The specific research and workshops on which this book is based were made possible by funding from the European Commission with additional support from Conoco Limited and Vauxhall Motors.

None of these contributors are responsible for the content of the book or the views expressed in it.

The New Economy of Oil
Impacts on Business, Geopolitics and Society

John Mitchell
with Koji Morita, Norman Selley and Jonathan Stern

First published in the UK in 2001 by
Royal Institute of International Affairs, 10 St James's Square, London SW1Y 4LE
(Charity Registration No. 208 223)
and
Earthscan, 8–12 Camden High Street, London, NW1 0JH

Reprinted 2005

Distributed in North America by
The Brookings Institution, 1775 Massachusetts Avenue NW,
Washington, DC 20036-2188

A catalogue record for this book is available from the British Library.

ISBN 1 85383 796 2 paperback
1 85383 745 8 hardback

Earthscan is an imprint of James & James (Science Publishers) Ltd and publishes in association with the International Institute of Environment and Development.

Typeset by Composition & Design Services, Minsk, Belarus
Printed and bound in the UK by Creative Print and Design
Original cover design by Visible Edge
Cover by Yvonne Booth

Contents

List of figures, tables and boxes

Figures

Tables

Boxes

Foreword

by Robert Mabro

Is our world subject to sudden and radical transformations? Are there discontinuities that change, in a significant way, the course of historical developments? Or, on the contrary, should we agree with the lament of Ecclesiastes: 'There is nothing new under the sun'?

Most people, no doubt, will agree that our modern world is in constant flux. Rare are the societies today, which can remain immune to external influences. They will always respond, either by adopting something new or by retreating deeper into their traditional culture. In both cases there is change.

Yet continuing changes do not necessarily imply discontinuities. Some 25 centuries ago, Heraclitus famously said that one cannot swim twice in the same river. *Panta rei*. This does not mean, however, that what is new (the river's water flow) is independent from what always remain the same (the river itself). And this is where the problem lies. How does one separate the new elements of an evolving picture from the constants that often constitute its main features? To focus on what is new is, of course, a very attractive option. I am not referring here to the media's obsession with news, or to those political spin doctors who try to disguise old wares under a label claiming that all that is on offer is new. I am thinking of serious work that takes a forward look, considers the future, and, by necessity, seeks to identify directions of change. Then, talking about what is new is of the essence. Yet one should never forget that continuities, far too often, tend to dominate. It may be boring to talk about things that have given the picture its familiar physiognomy, things that everybody claims to know and understand. Nothing turns people off more immediately than a sense of *déjà vu*. Nevertheless, the important point that readers should always keep in mind is that changes can be properly understood only in relation to continuities.

Sometimes I wonder whether historians always fully grasp this point. They often seem to indulge in a game in which one of them will emphasize the novelty of some important event, such as a revolution, only to be followed a few years later by a revisionist who will stress continuities, arguing that nothing significant has occurred. Well, this may be of great help in fostering academic careers. It is intellectually more honest, however, to hold the balance between what is emerging and the roots which give rise to new developments and often survive them for a very long time. Not very exciting perhaps, but the truth, if it is to be found, is more likely to lie in this balance.

John Mitchell, in this excellent book, is looking at the new elements which may influence, or indeed determine, the shape of the oil industry, the behaviour of the main actors, the forces of supply and demand, and the price path. And his forward look is well anchored in the constants which have characterized developments in the past 20 or 30 years. His work is not one of pure speculation. Mitchell does not consider the future as heralding a new beginning free from all the baggage carried over from the past. He carefully places the new elements in context, and thus draws a coherent picture which highlights both *nova et vetera*.

The New Economy of Oil is about a world where social acceptability is an increasingly important issue. People worry more and more about their health. But can we expect the trends that are being observed in smoking or in meat consumption, for example, to emerge in the oil domain? The issue relates to the trade-off involved. To refrain from smoking requires efforts to fight addiction. To avoid eating meat is easy so long as fish, poultry or pulses are available. But to give up the car is another story altogether. The trade-off is between the harm I am doing to the environment, harm that is caused by the multitude of which I am only an infinitesimal part, and my freedom. There is a difference in kind between our marriage to the car and addiction to smoking or meat consumption.

For social acceptability to make a difference, it will have to acquire the momentum of a considerable political force. Then governments may be persuaded to introduce all sorts of restrictions on the use of the car (to do so now would be political suicide), to impose technological changes on both refiners and car manufacturers (something is already happening on that front, but there is still a long way to go), and to invest in and subsidize public transport (which they hate to do). The issue will no doubt gain in salience, but not as soon as environmentalists may wish.

According to Mitchell, the availability of oil and gas resources is unlikely to be a major factor – as it would be if geology and technology were the only determinants of availability. The critical determinant is investment in productive capacity; and the problem there is that investment never corresponds to the requirements of changes in demand. Now there is surplus capacity, now a capacity constraint. This cyclical aspect is not likely to disappear.

I do not know how prices will move in the next 20 years. The oil/gas price relationship may well be inverted so that gas leads instead of oil. But then, which gas price? That which rules in North America, or in Europe? Or the cost of the marginal LNG tonne required by Korea or Japan?

Although it is now fashionable to refer to a world 'beyond petroleum', Mitchell wisely avoids spin. *The New Economy of Oil* is not about a post-petroleum world, but about the two decades ahead when oil and gas will be very much with us. New perceptions, yes. New structures, to some extent. New worries mixed with old ones about security of supplies, environment and cartelization, certainly yes.

That this book raises a host of issues for debate and further thoughts must have now become evident, although this Foreword has touched on only a very small number. The time has now come for the reader to discover and think about all of them.

Robert Mabro

Acknowledgments and disclaimers

This book is the final product of the '2020' Project which was financially supported by the European Commission and the companies listed on page ii. Much of the analysis and debate on which this book is based were generated in four workshops held under 'Chatham House rules' at the Royal Institute of International Affairs in 1999–2000, on the subjects of 'Oil Resources', 'Fuel for Transport', 'Gas for Oil Markets' and 'Strategic Challenges'.

The authors owe more gratitude than can easily be expressed to the vigorous arguments of the 130 people who came to the workshops from business, government, NGOs and institutes, from the Americas, China, Europe, Japan, the Middle East and Russia. The book does not pretend to summarize or represent their views.

The workshops benefited enormously from the technological papers and insights provided by Edouard Freund and his colleagues from l'Institut Français du Pétrole. Norman Selley was responsible for organizing three of the workshops and all were supported by Energy and Environment Programme staff: Ben Coles, Kate Kinsman and Matthew Thomas.

Many individuals contributed outside the workshops by commenting on draft chapters or by discussing key ideas: in particular, Thomas Ahlbrandt, Dennis Anderson, Fatih Birol, Jean-René Bauquis, Jean-Marie Bourdaire, Ronald Charpentier, Loren Cox, Denny Ellerman, Sylvie Cournot-Gandolphe, Patrick Criqui, Dominique Finon, Dermot Gately, Martha Harris, Henry Jacoby, Walid Khadduri, Wil Kohl, Ken Koyama, Knut Kuebler, Akira Miyamoto, Peter Odell, Robert Skinner and Eugene Skolnikoff. They too are not responsible for the final text.

Production of the book would not have been possible without the expert work of Margaret May and Matthew Link, of the Chatham House Publications Department; Gillian Bromley, the copy-editor; and Pauline Cowderoy, who assisted John Mitchell in bringing the texts into legibility and order.

The authors, like the participants in the workshops, did not try to reach consensus. The final responsibility for errors, omissions and future regrets lies with John Mitchell.

John Mitchell
Koji Morita
Norman Selley
Jonathan Stern

About the authors

John Mitchell is Chairman of the Energy and Environment Programme and Associate Research Fellow at the Royal Institute of International Affairs, London. He is also Research Adviser to the Oxford Institute for Energy Studies. From 1966 to 1993, Mitchell worked for British Petroleum, where his roles included Special Adviser to the Managing Directors, Regional Co-ordinator for BP's subsidiaries in the Western Hemisphere, and Head of BP's Policy Review Unit. His publications include 'Ethics and International Business', in *Annual Review of Energy and Environment*, 1999; as editor, *Companies in a World of Conflict* (London: RIIA, 1998); and *The New Geopolitics of Energy* (London: RIIA, 1996).

Koji Morita is a visiting Research Fellow with the Energy and Environment Programme at the Royal Institute of International Affairs. Prior to joining Chatham House in July 1999, he headed the Natural Gas Group at the Institute of Energy Economics, Japan. As well as working on this publication, during his time at the Institute Koji Morita has published a briefing paper, *Gas for Oil Markets* (London: RIIA, February 2000).

Norman Selley worked on the '2020' Project while Senior Research Fellow and Deputy Head of the Energy and Environment Programme at the Royal Institute of International Affairs. During his time at the Institute, he published a briefing paper entitled *Changing Oil* (London: RIIA, January 2000). Prior to joining Chatham House he spent 20 years in the oil and gas industry, with Conoco and Lasmo, where he headed the group's Economics Department between 1994 and 1999. He is currently Senior Economist with Amerada Hess.

Jonathan Stern is a London-based independent researcher and consultant specializing in natural gas issues. He is an Associate Fellow at the Energy and Environment Programme at the Royal Institute of International Affairs. He is also Senior Adviser at Gas Strategies. Publications include 'Soviet and Russian Gas: The Origins and Evolution of Gazprom's Export Strategy', in Robert Mabro and Ian Wybrew-Bond, eds, *Gas to Europe: The Strategies of the Four Major Suppliers* (Oxford: Oxford University Press, 1999); *Competition and Liberalization in European Gas Markets: A Diversity of Models* (London: RIIA, 1998); and *The Russian Natural Gas Bubble* (London: RIIA, 1995). From 1985 to 1992 he was Head of the Energy and Environment Programme and from 1990 to 1991 Director of Studies at the Royal Institute of International Affairs.

Units and conversions

Note on units

In this book we have tried to use the units that are customary for the subject under discussion. Thus money is counted in dollars, oil in barrels, natural gas in cubic metres (except for US readers, for whom a conversion is given below to discontinued British measures such as cubic feet, gallons and British thermal units). Tonnes are metric tonnes. Since this is a book mainly about oil, when fuels are compared they are compared on the basis of their oil equivalent energy content. For the same reason, the energy equivalent of hydro- and nuclear-generated electricity is given on an input basis (the amount of fuel oil needed to generate the same amount of electricity in a steam generator of average efficiency). Where we quote primary sources that use an output equivalent we have adjusted the primary energy numbers (and fuel shares) accordingly, so that primary energy numbers we quote from the BP Amoco *Statistical Review* and the IEA *World Energy Outlook* will be slightly higher than those in the original source.

Note on conversions

1999$ represents $ figures that have been adjusted to 1999 purchasing power by the US GDP deflator. At average 1999 exchange rates, $1 = €0.938, ¥113.91 and £0.618.

1 tonne of oil equivalent ≅

7.33 barrels of oil
1,111 cubic metres of natural gas
39,200 cubic feet of natural gas
1.5 tonnes of hard coal
0.805 tonnes of LNG
40.4 British thermal units
4,000 gigawatt-hours of electricity in a modern thermal power station

(Conversion factors from BP Amoco *Statistical Review*.)

Acronyms and abbreviations

ACEA	Association of European Automobile Manufacturers
APEC	Asia–Pacific Economic Cooperation Forum
APERC	Asia–Pacific Energy Research Centre
bbl	barrel = 159 litres
b/d	barrels per day
bn	billion ($=10^9$)
bn M^3	billion cubic metres
CAFE	Corporate Automobile Fuel Efficiency Standard (US)
CCGT	combined cycle gas turbine
CDM	Clean Development Mechanism
CFCs	chlorofluorocarbons
CITES	Convention on International Trade in Endangered Species of Wild Flora and Fauna
CNG	compressed natural gas
CO_2	carbon dioxide
cu ft	cubic feet (1 cu ft = 0.02832 cubic metres)
EC	European Commission
EIA	Energy Information Agency of the US Department of Energy
FSU	former Soviet Union
GATT	General Agreement on Tariffs and Trade
Gb	billion barrels
GDP	gross domestic product
GHG	greenhouse gases
IEA	International Energy Agency
IEO	*International Energy Outlook* (EIA)
IPCC	Intergovernmental Panel on Climate Change
IPP	independent power producer
JI	Joint Implementation
LNG	liquefied natural gas
LPG	liquefied petroleum gas
m b/d	million barrels per day
MEA	multilateral environmental agreement
mm Btu	million British thermal units (US). 1mm Btu = 293 kilowatt hours

mtoe	million tonnes of oil equivalent
M^3	cubic metre
m M^3	million cubic metres
NGL	natural gas liquids
NGO	non-governmental organization
NGPA	Natural Gas Policy Act 1978 (US)
NO_x	nitrogen oxides
NO_2	nitrogen dioxide
NYMEX	New York Mercantile Exchange
ODS	ozone-depleting substances
OECD	Organization for Economic Cooperation and Development
OPEC	Organization of Petroleum Exporting Countries
PEM	polymer electrolyte membrane
PNGV	Partnership for a New Generation of Vehicles (US)
PPP	purchasing power parity
PSA	production-sharing agreement
RF	recovery factor
SO_x	sulphur oxides
SO_2	sulphur dioxide
trn M^3	trillion (10^{12}) cubic metres
UNEP	United Nations Environment Programme
UNFCCC	United Nations Framework Convention on Climate Change
USg	US gallon (English wine gallon) = 3.785 litres
USGS	United States Geological Survey
VOCs	volatile organic compounds
WBCSD	World Business Council for Sustainable Development
WEO	*World Energy Outlook* (IEA)
WTO	World Trade Organization
ZEVs	Zero Emission Vehicles

Chapter 1

Overview

This book is for people involved with oil as consumers, competitors, commentators, investors, managers, politicians and regulators. It examines corporate, social and environmental challenges and argues that the chief concern about oil over the next 20 years will be its acceptability, not its availability. The book reviews conventional energy projections, focusing on issues of transportation, oil reserves and growing markets for gas. It argues that there is unlikely to be any sustained increase in fossil fuel prices over this period. Traditional 'oil' issues such as security are giving way to 'new' issues such as environmental and social behaviour in a context of changes in technology and increasing the globalization of competitive markets for all types of energy. The resulting challenges for the industries concerned, for national governments and for value-driven non-government groups are identified.

The key messages of the chapters are as follows:

1 *Overview:* So much change deserves a new perspective.
2 *The conventional vision:* Public projections of the role of oil are analysed in the benign scenario of continuing strong economic growth over the coming 20 years with some loss of market share to natural gas.
3 *Oil supply:* The relative abundance of oil and its uneven distribution form the basis for continued competition among producers. An aggressive oil price cartel cannot be sustained by exporting governments.
4 *Transport in transition:* Competition among alternative technologies for road transport is likely to intensify as 2020 approaches. New choices of transport fuel will be involved in the competition between vehicle manufacturers; new approaches to the offer of transport services will be tried.

5 *Gas for oil markets:* Gas is the new frontier for oil consumers and private sector oil companies. Growth is uncertain in many developing countries because of the need to invest in cross-border infrastructures. There is a possible contradiction in the conventional vision of gas gaining a greater share of energy markets across the world while it also increases in price relative to oil.
6 *Oil prices – the elastic band:* Competition will force oil prices to fluctuate between a lower limit determined by exporters' resistance to economic disaster and a ceiling which is kept down by competition from other fuels and technologies of demand.
7 *Energy security:* Objectives and instruments need to be redefined: the key *economic* concerns for importers are managing the effects of disruptions and minimizing long-term energy costs, not minimizing their energy imports. The threat of *political* sanctions has been reversed. It is exporting countries that now need to be concerned about sanctions aimed at their domestic and foreign policies by governments and public opinion in developed countries.
8 *Environment and social acceptability:* The Kyoto Protocol is a step towards internationally agreed policies to limit the growth of fossil fuel consumption but its quantitative approach, obsolescent baselines and limited geographic scope mean that other measures will be necessary. Upstream, energy projects face increased opposition through private international channels to limit the ecological, social and political damage caused by high-impact development projects in poor or weak countries.
9 *Challenges and choices:* The risks and choices for companies, governments and advocates of social values are different, although they are interconnected. They are differently affected by the risks in the 'conventional vision' for oil, and by conflicts over the acceptability of oil use and development. Understanding the differences is the first step to identifying the scope for mutually interesting actions.

Chapter 2

The conventional vision

This chapter analyses the current (mid-2000) 'business as usual' or 'reference case' scenarios of energy and oil trends to 2020 of the three main agencies that expose their projections – and their methods – to public scrutiny.[1] These are:

- the European Commission (EC), through its 1999 'Shared Analysis' based on the POLES model of the Institut d'Economie et de Politique de l'Energie, referred to here as EC (POLES);
- the Energy Information Agency (EIA) of the US Department of Energy, in its annual *International Energy Outlook*, referred to here as EIA (*IEO*-00) or EIA 00; and
- the International Energy Agency (IEA), through its biennial *World Energy Outlook*, referred to here as IEA (*WEO*-98) or IEA 98. A new *World Energy Outlook* was due for publication in November 2000.

The three agencies tell similar stories. Despite different dates and methods, they can be regarded as samples of the same vision of what would happen under policies, consumer and citizen demands, and business strategies which continue past trends.[2] This 'conventional vision' does not assume implementation of the commitments entered into under the

---

[1] European Commission, 'European Energy Outlook to 2020', in *Energy in Europe*, special issue, December 1999 (Brussels; the Shared Analysis Project working papers (SAP:WP) are available on <www.shared-analysis.fhg.de> and from the Fraunhofer Institute for Systems and Innovation Research, Karlsruhe, Germany); EIA, US Department of Energy, *International Energy Outlook 2000*, April 2000 (Washington DC); IEA/OECD, *World Energy Outlook*, November 1998 (Paris; the new IEA *World Energy Outlook* is due for publication in November 2000).

[2] The forecasts contain comparisons that support this similarity. See EIA *IEO-00*, pp. 15–20; EC (POLES) Shared Analysis Project, *Economic Foundations for Energy Policy*, vol. 2 (Grenoble: Laboratoire du Centre National de la Recherche Scientifique, 1999), pp. 40–6.

Kyoto Protocol, first because the protocol was not ratified at the time the projections were made, and second because we do not yet know either what measures may be taken to implement the protocol or what their effects may be.[3] There is likely to be less agreement about these measures and their effects.[4] Nevertheless, the trends of the 'conventional vision' are the basis for discussions of serious climate change mitigation policies which would alter these trends were they widely adopted.

There are differences among the projections – mainly at the regional and sectoral levels – which can be regarded as sensitivities rather than alternative projections. Each study illustrates variations in economic growth, oil prices and energy intensity. Close comparisons among the studies at a regional or national level are difficult because of differences in definition, while generalizations across countries and regions are also dangerous, since local circumstances affect demand and supply by means other than the international prices or the energy aggregates.

This book approaches the conventional vision from a particular perspective. Our question is whether the energy trends described in the 'reference' cases contain such uncertainties, discontinuities and tensions that actions taken to resolve them by governments, private sector companies, and even consumers and citizens collectively may themselves change the energy game. What factors can we see that would alter 'business as usual' *before* we consider the impact of serious climate change mitigation policies? Such alterations would be of immediate interest to those responsible for business, policy and the advocacy of

[3] The UNFCC includes a general commitment by OECD members and certain other 'Annex 1' countries to reduce their emissions of greenhouse gases to 1990 levels. The Kyoto Protocol sets specific national targets for reducing an aggregate of greenhouse gases and offsets at various points during the period 2008–12; for a description of the commitments see Michael Grubb, Duncan Brack and Christiaan Vrolijk, *The Kyoto Protocol* (London: Earthscan/Royal Institute of International Affairs, 1999). The three projections discuss targets in terms of CO_2 emissions only (EIA: pp.153–66; EC (POLES 1): ch. 4, pp. 68–99; IEA: p. 24), but EIA (*IEO*-00) provides a historic database of methane and NO_x emissions.

[4] For a recent review, see Ulritsch Bartsch and Benito Müller with Asbjørn Aaheim, *Fossil Fuels in a Changing Climate* (Oxford: Oxford University Press/Oxford Institute for Energy Studies, 2000), esp. ch. 16, pp. 297–310.

values, whatever the form and severity of future climate change mitigation policies. They are also relevant to the design of such policies in an effective and cheap form, for climate policies designed to change trends may turn out to be unexpectedly costly or even useless if those trends have changed already.

The conventional starting point: energy and GDP

Conventional projections of demand for commercial energy are associated with projections of Gross domestic product (GDP).[5] (GDP is not the only useful measure of development, but it is the one generally associated with energy projections.) The conventional procedure, used in all three scenarios under discussion, is to project population growth, growth of GDP per head and energy intensity of GDP. In the conventional vision, world primary energy consumption would grow by about two-thirds between 1995 and 2020, in a world economy which would double in size. Even at the global level, these projections contain uncertainties.

First, some of these projections, made in 1998–2000, may not allow fully for the effect of low growth of the economy and energy consumption in the period 1997–2000. World primary energy consumption actually fell in 1998.[6] The average annual growth in global primary energy demand for 1995–2000 is likely to be nearer 1% than the 2% on which the conventional visions are based; and three years' lost growth of 1% at the beginning, compounded for 20 years, will lower the 2020 figure for global energy consumption by more than 10%.[7] The EIA and EC (POLES) cases allow for a 2–3-year interruption in growth in Asia, but then pull back to the trend line. The

[5] The IEA projections show estimates of non-commercial energy for the OECD countries, but it is in the non-OECD countries that these are likely to be a significant fraction of total energy use: 24% of total Indian energy (but only 6% of Chinese energy consumption), according to the World Bank, *Development Indicators 1999*, table 3.8.

[6] BP Amoco *Statistical Review 1999*, p. 38.

[7] See discussion in EC (POLES 3), section 1.2, pp. 21–6. The EC (POLES) reference case for 2020 shows Asian demand of 11.9% compared to a non-crisis situation. In its short update of *World Energy Outlook 1999*, *1999 Insights* (published in March 1999), the IEA appears to have made similar reductions compared to EIA, *International Energy Outlook 1998*.

projected annual increase of consumption by 5,300–6,500 million tonnes of oil equivalent (mtoe) per year worldwide compares with a recent history of much lower increases: approximately 3,000 mtoe annually over 1973–98.

Second, the projections cited already have a range of 1,200 mtoe in their energy projections for 2020. This is more than the current production of Middle East oil. About half of this range can be explained at a global level by slightly different assumptions about trends in economic growth and energy intensity (see Table 2.1). At growth rates of around 3%, a difference of 0.1% in the growth rate of either GDP or energy over 25 years makes a difference of about 5% in the projection for the final year.

Table 2.1: GDP growth and energy intensity comparisons (%)

	EC (POLES)	EIA (*IEO-00*)	IEA (*WEO-98*)
GDP growth	3.35	2.9	3.1
Energy intensity	–1.2	–0.8	–1.1
Primary energy growth	2.15	2.1	2.0

By rough adjustments, Figure 2.1 shows the effect of adjusting, at the global level, to the 'round numbers' of 3% GDP growth and 2% primary energy growth which are broadly consistent with past trends.[8] The effect is roughly to halve the range of the forecasts.

Third, all these conventional visions were built from regional or national projections, all project GDP rising more rapidly in developing countries than in industrial countries (and rising very slowly in the emerging market economy of the former Soviet Union). The effect would be that between 1997 and 2020 the developing countries' share of world GDP would increase by about 10%, with a corresponding fall in the share of the industrialized countries. It is difficult to be certain whether this shift would increase the average energy intensity of the world economy.

Developing countries generally appear to have more energy-intensive economies. In 1996 $1 of Chinese GDP, measured at current US$, required 1,430 g of energy oil equivalent, while $1 of US GDP required

[8] The numbers used in the projections and adjustments for IEA 98 have been adjusted to represent hydropower on a primary fuel input equivalent basis.

Figure 2.1: Effect of energy intensity and GDP projections

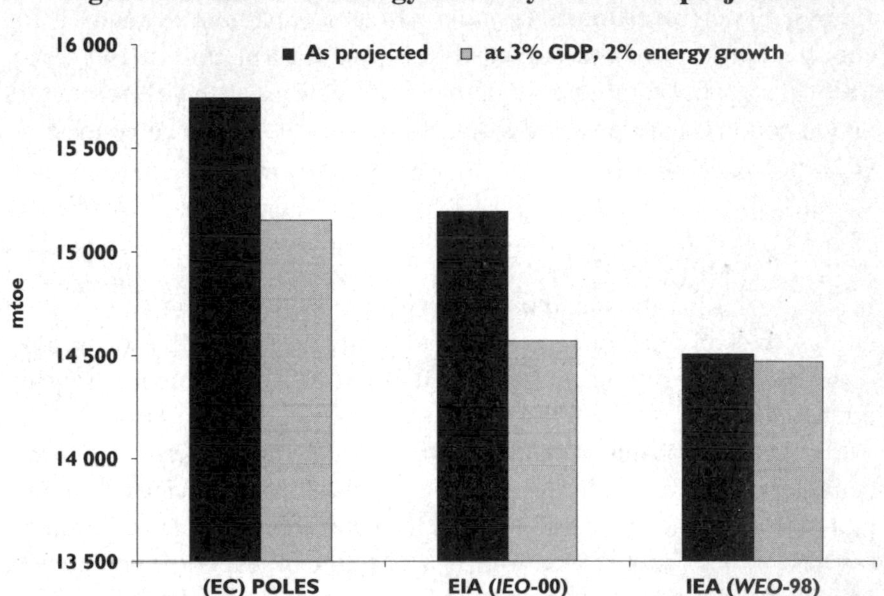

only 290 g.[9] However, between 1980 and 1996 the energy/$ of GDP ratio fell by 22% in the US; in China the fall was 57%.[10] In the Chinese case the improvement appears to have been mainly due to improvements in technical efficiency: structural changes alone would have tended to increase energy intensity through the rapid development of energy-intensive industrial sectors.[11] Presumably the technological improvements in China were largely the result of Chinese economic reforms which enabled Chinese enterprises to introduce technology

[9] World Bank *Development Indicators 1999*, table 3.8, pp. 148–9.

[10] GDP is difficult to compare across countries because exchange rates fluctuate. If exchange rates are restated on the basis of 'purchasing power parity' (PPP) to reflect the power of the currency to purchase similar goods, the GDP of many developing countries appears higher, because prices are lower in developing countries for goods and services which are not traded. On a PPP basis, energy intensities in China and the United States appear closer: 230 g per PPP$ in China and 261 g per PPP$ in the United States in 1997. The improvements still differ: from between 1987 and 1997 China reduced its energy intensity by 52%, and the US by 27%.

[11] See Richard Garbaccio, Mun S. Ho and Dale Jorgenson, 'Why has the Energy–Output Ratio Fallen in China?' *The Energy Journal*, vol. 20, no. 3 (1999), pp. 63–91.

already available globally, plus a direct transfer of technology through foreign direct investment. (Foreign direct investment accounted for about a quarter of Chinese gross fixed capital formation in 1995.[12])

For the global total energy demand projections it therefore matters which countries are growing faster. However, the evidence connecting GDP growth and energy demand is not simple.

Energy and development

It does seem that the obvious is true: people in rich countries use more energy than in poor ones, whether to generate more GDP or to consume more energy-related services. It is not clear that this relationship is changing.

Figure 2.2 compares per capita GDP with per capita energy use for 32 countries. These exclude the countries of the former Soviet Union and eastern Europe, for which it is difficult to construct PPP data; Germany, because of the lack of PPP data for 1987; and the oil-exporting countries of Organization of Petroleum Exporting Countries (OPEC), Mexico and Norway.[13] There is little difference between 1987 and 1997 in the relationship between energy consumption per capita and GDP per capita: both increase together. The United States and China are both somewhat above trend, possibly because of subsidized energy prices in China and the lack of energy consumption taxes in the United States, compared to most Organization for Economic Cooperation and Development (OECD) countries.

The messages are clear:

- Higher GDP per head and higher energy consumption per head are strongly linked.
- There are significant differences between countries.[14] It is these differences that are the subject of econometric analysis to probe the

[12] UNCTAD, *World Investment Report 1997*, fig. II.18, p. 82.

[13] The United States, Canada, Argentina, Brazil, Chile, Austria, Belgium, Luxembourg, Denmark, Finland, France, Greece, Ireland, Italy, the Netherlands, Portugal, Spain, Sweden, Switzerland, Turkey, the UK, Egypt, South Africa, Australia, Bangladesh, China, India, Japan, New Zealand, Pakistan, Philippines, Singapore, South Korea and Thailand.

[14] The high energy intensity of Singapore is probably an aberration: either its apparent consumption figures do not adequately reflect oil re-exports, or it has an energy-intensive industrial structure attributable to its regional refining business.

Figure 2.2: Energy per capita versus GDP per capita

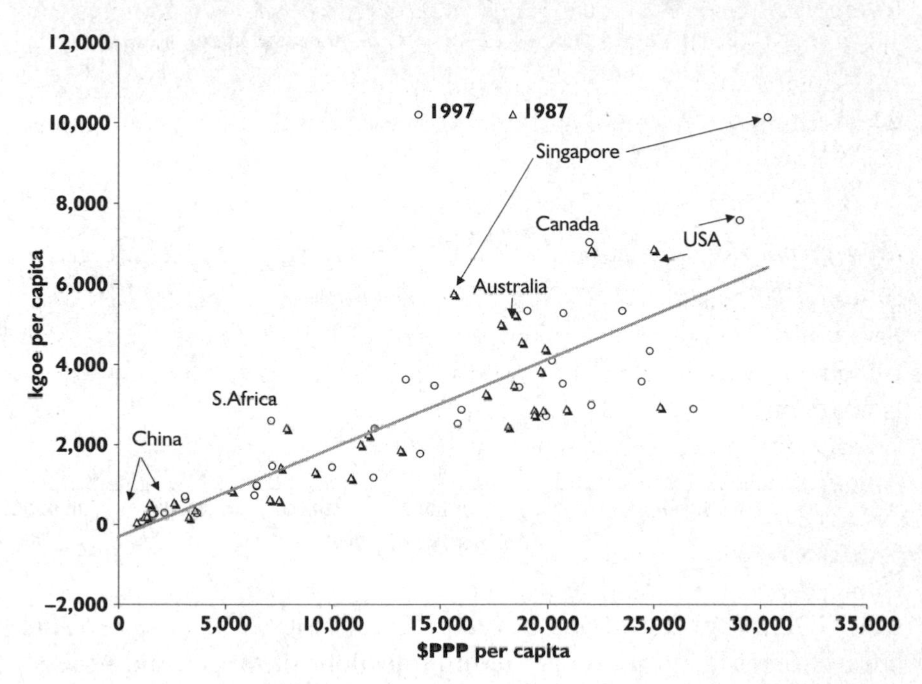

Note: $R^2 = 0.7216$.

underlying effects of differing levels of income, infrastructure, technological inheritance, consumer prices and taxes. The institutional and policy framework within which consumers choose to combine spending on energy with other categories of expenditure, and firms combine energy with inputs of capital, organization and labour, is also important, but is difficult to address statistically.

Figure 2.2 is a simple chart and does not show energy intensities changing over time as incomes rise. There is, however, some evidence that the economies with high incomes per head are less energy intensive. The data used in Figure 2.2 can be presented to illustrate this, as shown in Figure 2.3.

The idea of energy intensity falling with GDP growth is used in the conventional vision to project a steadily weakening link between energy demand (the global rule of thumb seems to be a 1% difference)

Figure 2.3: Energy/income elasticity versus level of income

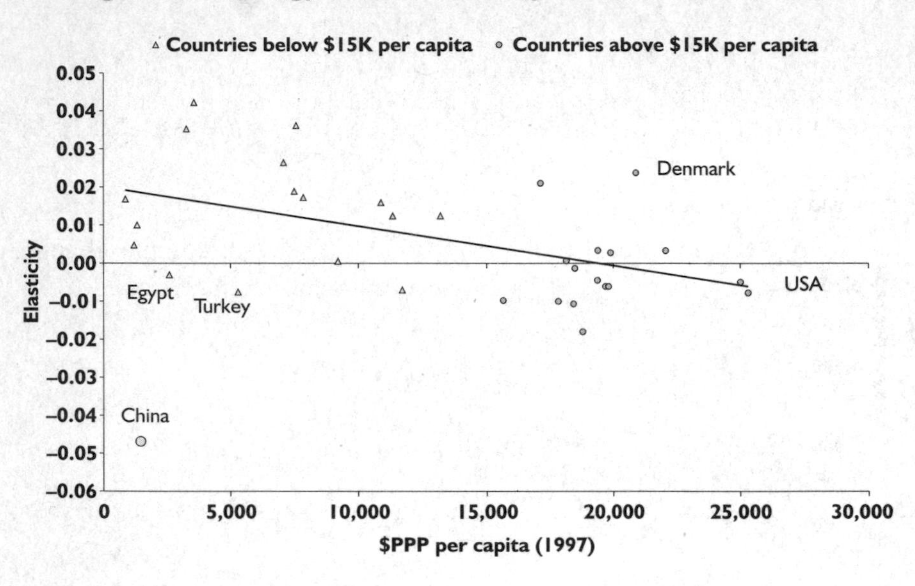

and GDP in developed and developing countries in the future. This idea is also used in the longer-term projections of the Second Assessment Report of the Intergovernmental Panel on Climate Change (IPCC).

Energy and GDP

Many studies seek to explain the national differences by analysing different activities, structure and intensity on the lines developed by Schipper and Myers:

$E = \sum A_k I_k$, where A = activity, I = intensity, and k = sector.[15]

With this formulation it is possible to separate out the effects on total energy consumption of changing industrial or consumption structure (k), the level of activity or growth (A) and the intensity of technical

[15] *The Link Between Energy and Human Activity* (Paris: IEA/OECD, 1997); Lee Schipper and Stephen Myers, *Energy Efficiency and Human Activity* (Cambridge: Cambridge University Press, 1992).

efficiency (*I*). This separation is used to some extent in the conventional vision projections,[16] but the activities are broadly defined as transportation, electricity and the rest; the last category is divided (in only EC (POLES)) between industrial and household energy use. Further subdivisions are available in national projections which the compilers of the international studies also use.[17] International projections which include a more detailed sectoral breakdown and the opportunity to trade between countries are available from models used in climate change policy studies for the (IPCC). In a review of models by the Energy Model Forum at Stanford University even the baseline projections of these models show a much wider variation – even for GDP assumptions – than exists in the 'conventional vision' of the three official studies discussed here.[18]

The separation of activity, structure and intensity may be useful in explaining current trends and identifying the possible impacts of specific policies – for example aimed at changing energy intensities – but the problem of projection remains. 'Business as usual' may be argued to include 'change as usual' since even at the sector level history has included changes.

There are several intuitive reasons why energy intensity should change as personal incomes and GDP rise. They would carry different weight in different countries at different times. Starting from low levels of income and economic development:

- Household consumption will change from being dominated by basic heat to rapidly rising energy use for higher levels of comfort in space heating and cooling (and larger dwellings), and greater use of electrical appliances, finally to a degree of saturation influenced by

---

[16] IEA, *World Energy Outlook* 1998, pp. 41–3 and figs. 3.8 and 3.11; EIA 00, p. 6 and fig.11.

[17] See e.g., for the United States, EIA, *Annual Energy Outlook 2000* (Washington DC: US Department of Energy, 2000); for the European Union, European Commission, 'European Union Energy Outlook to 2020'; for Japan, *Handbook of Energy and Economic Statistics '99* (Tokyo: Energy Data and Modelling Center, Institute of Energy Economics).

[18] John P. Weyant and Jennifer N. Hill, 'Introduction and Overview', in *Energy Journal*, special issue on *The Costs of the Kyoto Protocol: A Multi-Model Evaluation*, 1999, p. xxii and fig. 7. See also Bartsch and Mueller, *Fossil Fuels in a Changing Climate*, pp.118–21.

the income distribution patterns of the country concerned. Income distribution typically changes very slowly, so that the 'technical' market for heat will never be saturated because there will always be a proportion of poor people living in small spaces less comfortably than the average.

- Industrial energy consumption will be influenced by technical efficiency within each sector, and by changes in the structure of the economy, e.g. changing proportions of agriculture, heavy and light industry, and services. One may eventually see evidence of diminishing marginal returns to additional energy inputs compared to other inputs.[19]
- Energy consumption in the energy transformation sector may be influenced by income – which drives the demand for electricity to grow faster than the demand for heat – but is also subject to the chosen technology of transformation, which in turn is influenced by the cost and availability of primary energy inputs (fuels).

Data difficulties

Some studies suggest that the level of GDP per head beyond which energy use grows less fast than GDP is quite high.[20] However, energy is not the only input. Capital and labour and their respective prices and technical possibilities may also combine differently in different countries. Taking account of these, high energy input per unit of GDP may be the most efficient outcome.[21] Similar arguments may apply to the allocation of household budgets: the relative price of other items of domestic expenditure affects this, just as income affects the budget allocation. There are data difficulties in international comparisons, and

[19] See John R. Moroney, 'Output and Energy: An International Analysis', *Energy Journal*, vol. 10, no. 3 (1989).

[20] See Marzio Galeotti and Alessandro Lanzen, 'Richer and Cleaner? A Study of Carbon Dioxide Emissions in Developing Countries', *Energy Policy*, vol. 27, no. 1 (1999), pp. 561–73, who find that CO_2 emissions elasticity switches (depending on the model) at between $16,600 and $15,000 (1990$); $19,900 and $18,000 (1999$).

[21] Andrew Atkinson and Patrick J. Kehoe, 'Models of Energy Use: Putty–Putty versus Putty–Clay', *American Economic Review*, vol. 94, no. 4 (September 1999), pp.1028–43.

the choice of econometric model is important, but it does appear that the income elasticity of household demand for energy other than electricity (but including transport) falls with income and is less than 1 as incomes pass $7,000 (1999$) per capita.[22] The elasticity of household demand for electricity also falls with income but remains greater than 1 (1.4 in the study) even at incomes of $1,400 (1999$). These findings fit the intuition that there are saturation effects in household energy consumption.[23]

Sector demand

National variations in energy intensity across different sectors, in a large number of countries, over long periods of time, are addressed in a major study by Judson et al.[24] This study shows strong results for the effect of income levels on energy demand by the industry and construction, transportation and household sectors. Income explains over 80% of the national differences in these sectors, with important, but smaller, effects on demand for agriculture, the energy sector and non-energy uses. The results show strongly that as incomes rise, the related rise in energy demand is at first strong (high income elasticity) and then weakens, with the degree of weakening dependent on the sector (it is greatest in the 'household and other' sector and least in the transport sector). The turning points are lower than in the study cited in the previous paragraph: around $3,500 (1999$). Above $12,248 (1999$) the income elasticity of energy use appears to approach zero or go negative (although the authors doubt this interpretation) for the aggregate and all categories except transport, where it remains at 0.5.

---

22 Dale Rothman, J. Hong and T. N. Mount, 'Estimating Consumer Demand Using International Data: Theoretical and Policy Implications', *Energy Journal*, vol. 15, no. 2 (1994), pp. 67–88.

23 For a review of the problems of estimating elasticities, see Jago Atkinson and Neil Manning, 'A Survey of International Energy Elasticities', in T. Barker, P. Ekins and N. Johnstons, eds, *Global Warming and Energy Demand* (London and New York: Routledge, 1995), pp. 47–105.

24 Ruth Judson, Richard Schmalensee and Thomas Stoker, 'Economic Development and the Structure of Demand for Commercial Energy', *Energy Journal*, vol. 20, no. 2 (1999), pp. 29–58.

Income elasticities of 0.6–0.7 are consistent with the '3% GDP, 2% energy' rule of thumb used as the basis for Figure 2.2.

The same study also shows that country-specific factors (including income differences) account for more than 96% of national variations in total energy consumption.

Projections

Against this background it is interesting to review the three 'conventional vision' studies from the point of view of regions and sectors.

For the longer term, projection is even more difficult than for the next 20 years. The IPCC Second Assessment Report projections to 2050 have been seriously challenged by econometric studies which show that the improvements (reductions) in energy intensity projected by the IPCC in its 'business as usual' scenario are outside the bounds of what has been achieved historically with normal structural and technological change.[25]

Regional differences

Exact comparisons are not possible because of differences in the way regions and categories are defined by the three agencies under consideration,[26] but it seems that the Energy Information Agency of the US Department of Energy (EIA) GDP assumptions are higher than those of the IEA for North America,[27] China and developing countries, while the POLES assumptions are higher than both for developing countries and for the FSU and eastern Europe. The EIA has the highest assumption for China, although (surprisingly, given the uncertainties) the three projections are closer to one another for Asian industrial countries, where GDP

[25] R. Schmalensee, T. M. Stoker and R. A. Judson, 'World Carbon Dioxide Emissions, 1950–2050', *Review of Economics and Statistics*, vol. 80, no. 1 (1988), pp. 15–27.

[26] The EIA treats Mexico, but not Korea or Turkey, as an industrial country. IEA (98) treats Mexico and Korea as developing countries and excludes them from the OECD totals. In this chapter the IEA grouping is followed and the EIA numbers adjusted accordingly.

[27] The EIA 00 reference case growth rate for 1995–2020 for North America is 2.3%, compared to 2.7% in the IEA 98.

Figure 2.4: Energy use projections by region

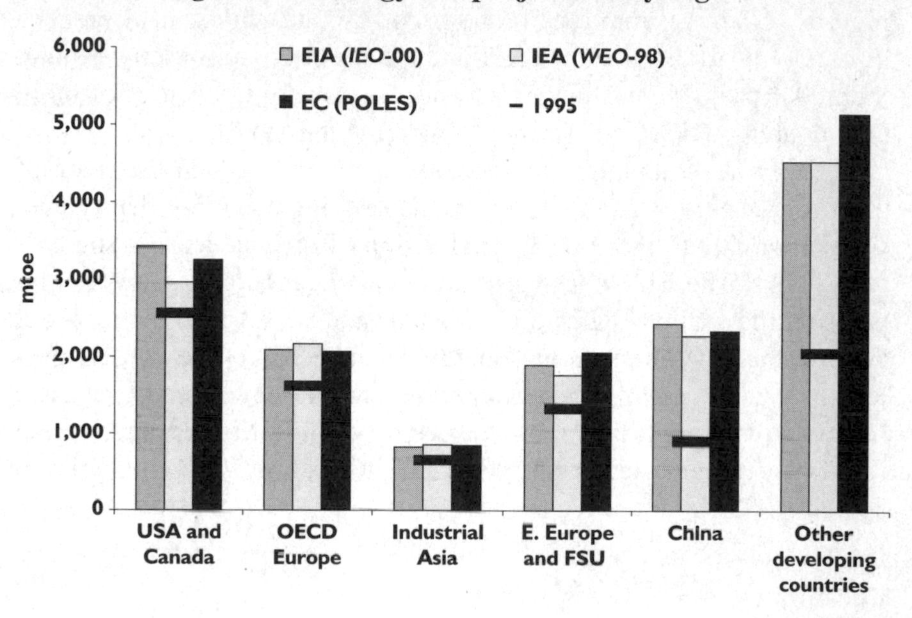

Figure 2.5: Regional shares of world energy use

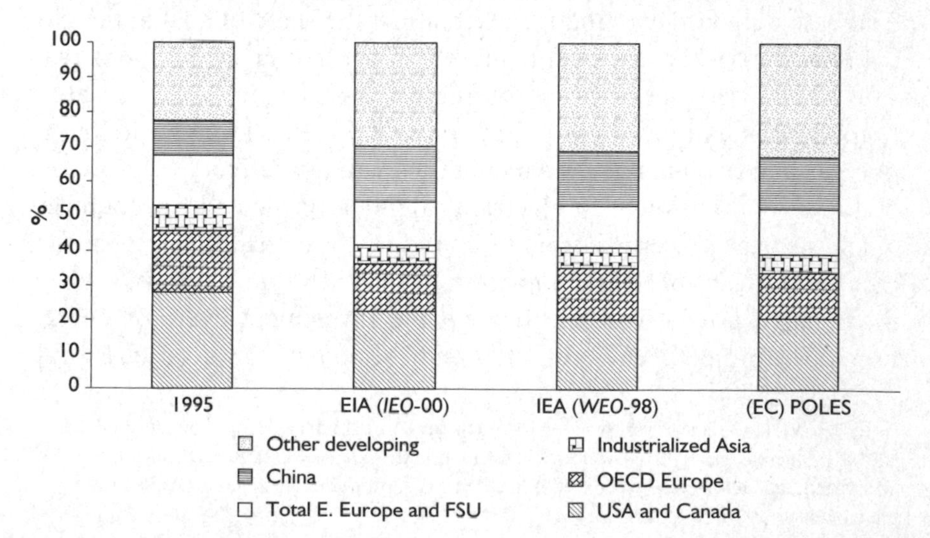

growth is projected at below the world average and below historic records. These assumptions broadly correspond with the differences in energy projections shown in Figure 2.4, but do not exactly explain them. All projections show energy use in developing countries more than doubling (GDP is projected to increase threefold).

These projections imply an increase of around 10% in the share of the world energy market held by the developing countries. This corresponds roughly to the conventional vision of their increasing share of world GDP. The EIA is less pessimistic about economic growth prospects in the United States and Canada, and more pessimistic about Europe, than the International Energy Agency (IEA). The POLES projection sees the greatest increase in the share of developing countries. In this projection, industrial countries (excluding Mexico and Korea) see their share of world energy use fall just below 40%, compared to 53% in 1995 (Figure 2.5).

Fuel mix

Differences in regional energy demand will affect the global fuel mix: high demand in North America, with strong demand for transport, will affect petroleum demand more than high demand in developing countries.[28] High electricity demand in countries with a competitive domestic coal base to their power industry will affect the share of coal in the global energy mix because coal is more competitive in such regions than in regions where coal has to be imported and there are alternative fuels domestically available. Constraints on maintaining or adding to nuclear power plants in certain countries will also affect the result, depending on the growth in electricity demand projected for countries where nuclear power is currently important. The change in global fuel mix is the result of the combination of these factors – and, of course, the relative prices working through the investment cycle of energy users. The projection systems attempt to simulate this in various ways.

[28] In this book, fuels are compared on the basis of their primary energy input equivalent. Inputs of nuclear and hydropower are therefore quantified at the fossil fuel replacement equivalent, not at the energy content of the electricity produced. IEA and POLES figures have been adjusted accordingly.

Figure 2.6: Fuel share of energy growth, compared to 1995

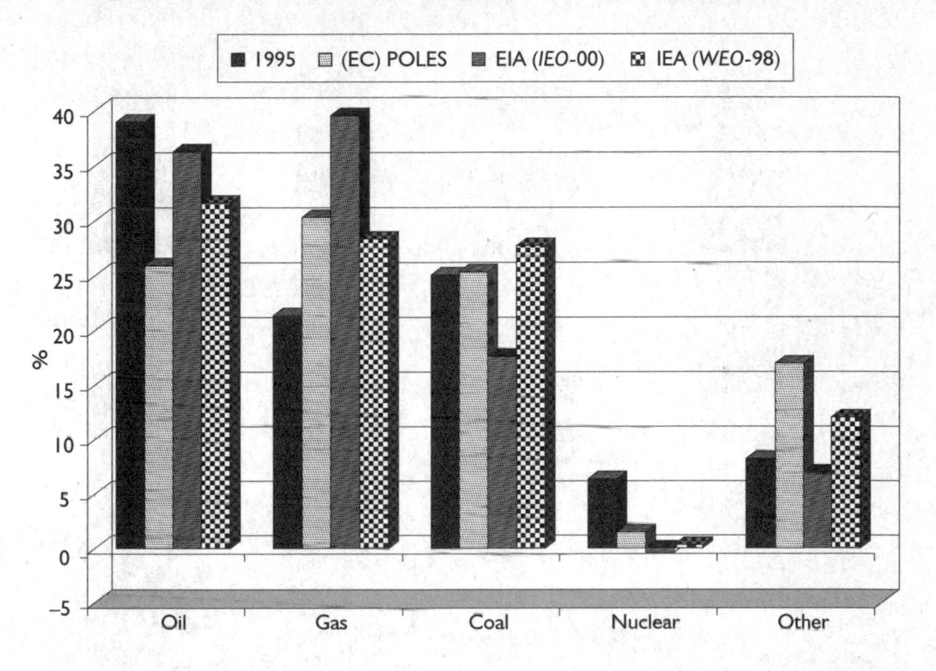

The global results are shown in Figure 2.6. For each fuel, the left-hand column shows the fuel share in 1995, while the three columns to the right show for each projection that fuel's share of the increase in world energy use. If a fuel's share of the increase is lower (higher) than its share of 1995 demand, this implies that its share of 2020 demand will be lower (higher) than in 1995. (The resulting total shares are shown later in Figure 2.7.) The difference between the total 1995 share and the share of the increase is an indicator of the structural change implied on both the demand and the supply side, and a measure of the shift in resources which would accompany the change. For nuclear and gas, in all projections, these shifts are large.

All the projections show gas gaining a higher share of growth than its present share, with oil losing most in the POLES and least in the EIA projection. This partly reflects the POLES prediction for relatively slightly lower economic growth in North America and relatively higher economic growth in developing countries, particularly Latin America, where the scope for nuclear, hydroelectricity and renewable

Figure 2.7: Fuel shares of world primary energy use, 1995 and 2020

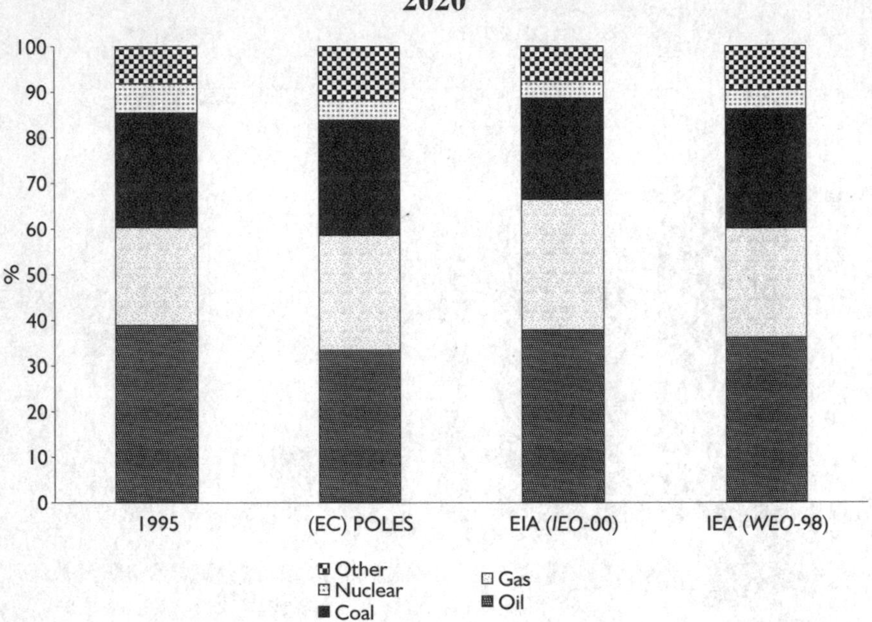

energy is supposed to be higher. Except in the EIA projection, coal is projected to increase its market share (partly driven by the high growth forecast for developing countries with coal-based electricity). Only in the POLES projection would there be a large shift into 'other' hydro and renewables. Some of these shifts would be concentrated in particular industries and regions, as will be shown later.

The resulting shares of total primary energy use in 2020 are shown in Figure 2.7. In all the projections the oil share falls, but remains the largest single source of primary energy, between 35% and 40% of the total. The nuclear share falls in all cases to around 4% of the total: to retain its share would require nuclear supply about 75% above that projected. The gas share increases substantially in all projections, but the variation between the projections is large (4–8%, compared to a 1995 share of 21%). As will be shown later, these variations are concentrated in certain regions and markets and probably constitute one of the main uncertainties of the 'conventional vision'. In the EIA case the share of hydroelectricity and renewables actually falls. Only in the EIA projection does the coal share fall (or even change much), mainly as a result of the greater substitution of gas for coal in western Europe in that projection.

The projections differ on where the changes in fuel shares occur. The biggest differences show up in the changes in fuel share for natural gas in developing countries. The EIA projection is more optimistic for oil and pessimistic for coal, maintaining their shares in industrial country markets. Differences in share changes for nuclear in developing countries are changes in very small shares.

Table 2.2: Changes in fuel share of primary energy consumption, 1995–2020 (%)

	EC (POLES)	EIA (*IEO-00*)	IEA (*WEO-98*)
Oil			
Industrialized countries	−6	−1	−3
Developing countries	−6	−2	−4
World	−5	−1	−3
Gas			
Industrialized countries	3	7	−1
Developing countries	6	10	5
World	4	7	3
Coal			
Industrialized countries	2	−3	9
Developing countries	−4	−5	−4
World	0	−3	1
Nuclear			
Industrialized countries	−1	−4	−4
Developing countries	0	−2	2
World	−2	−3	−2
Other			
Industrialized countries	3	1	−2
Developing countries	4	−2	2
World	4	0	1

Sector use: the electricity conveyor

Fuels do not compete in all sectors; for example, the transport sector is dominated by oil. Nuclear and hydroelectric power (and most renewables) reach the user through electricity; electricity itself competes with the direct burning of fossil fuels. Electricity provides the means by which other fuels can compete with oil and gas in sectors such as space heating and process heat. It also is the only means of powering applications such as motors, computers and lighting: these subsectors are difficult to analyse, except in detailed national projections such as the EIA Annual Energy Outlook.[29] However, there is strong evidence that higher incomes do not weaken the demand for electricity so much as the demand for energy in total (in contrast to the effect on the demand for non-electric energy forms).[30]

Table 2.3: Growth in final energy use by fuel and sector, United States, 1973–98 (%)

Fuel	Domestic and commercial	Industrial
Coal	–44	–43
Gas	2	–4
Petroleum	–52	–1
Electricity (final use)	111	53
Electricity (inc. generating and transmission losses)	95	41
Total (final use)	11	–4
Total (inc. generating and transmission losses)	38	8

Source: US Annual Energy Review, Table 2.1. Note that totals include hydro and renewables; electricity generating and transmission losses are on net purchases from utilities. There is a discontinuity between 1989 and 1990.

Final demand for energy in the US domestic and commercial sector increased only 11% between 1973 and 1998,while personal consumption expenditure increased nearly sevenfold in constant dollars.[31] In US industry, final demand for energy actually fell by 4% between 1973 and 1998, while industrial output doubled.[32] But the demand in both sectors was turning to electricity. Final demand for electricity doubled in the domestic and commercial sector and increased by over 50% in the industrial sector. This increased the energy needed to meet the final demand because generating and transmitting electricity uses energy.

[29] EIA, US Department of Energy, *Annual Energy Outlook 2000* (Washington DC).
[30] Dale Rothman, J. Hong and T. Mount, 'Estimating Consumer Demand Using International Data', *Energy Journal*, vol. 15, no. 2 (1994), pp. 67–88, esp. table 3, p. 85.
[31] US Bureau of Census, *Statistical Abstract of the United States 1999*, Table 1434.
[32] Ibid., table 1438.

Figure 2.8: Primary fuels used to supply US industrial energy demand, 1973–98

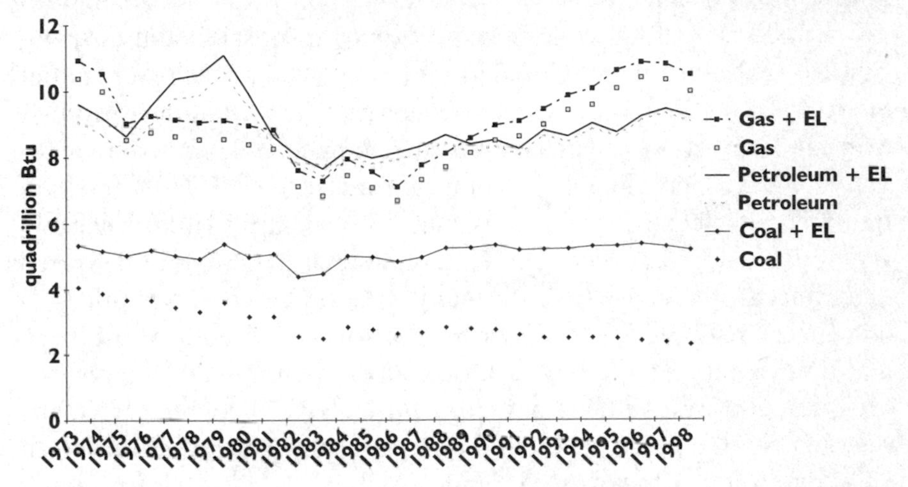

Source: Adapted from EIA, *Annual Review of Energy 2000*, Table 2.1.

However, that increase was mitigated by changes in the sources of electricity (the nuclear share increased by 16%), so that the energy needed to meet the electricity demand increased slightly less than the demand for electricity itself in final use. Nevertheless, the net effect was an increase of 38% (rather than 11%) in energy required for the domestic sector and 8% (rather than a fall of 4%) in energy required to meet the final demand in the industrial sector (see Table 2.3).

It is through the growth in electricity demand that fossil fuel supplies of energy to the domestic, commercial and industrial sectors continue to increase. In the US electricity sector between 1973 and 1998, coal replaced gas and oil, supplying 56% of all electricity and 80% of the electricity generated from fossil fuel. The coal share of primary inputs to electricity increase by 10% and 17% respectively. Figure 2.8 shows how coal has, through electricity, effectively maintained its supply to the US industrial sector despite a steep decline in the demand for coal for final use. The curve 'Coal + EL' shows the final energy demand for coal, plus coal's share of the supply of fuel to meet the final demand for electricity.

New market structures for electricity

In most countries the organization of the electricity industry is undergoing profound change. Communications and control technologies allow the 'unbundling' of generation from transmission and distribution to final consumers. Competition is allowed and encouraged at every possible stage (and where it is not practical, as in the operation of local distribution systems, competing shippers are entitled by regulation to equal opportunities to compete for access).[33] These reforms in the electricity sector have in many industrial countries led to a narrowing of the margin between fuel input prices and the prices of delivered electricity to the consumers. In many developing countries prices to consumers had been held down by controls, or directly subsidized, and investment restricted by lack of funds in public sector power monopolies. Reforms in developing countries have a double effect: the opening of electricity generation to foreign investment has attracted an expansion of capacity,[34] but has also created a demand for higher prices, at least for purchases from the private sector plants, to provide a return on investment.[35]

There is also a technical dimension to competition among fuels for electricity generation. The fuels used have very different properties: hydroelectric and nuclear generators produce no carbon dioxide or methane. The thermal efficiency of combined cycle gas turbine generators is up to 50% greater than that of steam generators using other fossil fuels. The balance between capital and fuel costs is different, so that the economics involved in choosing what type of new plant to build are different from those involved in choosing which existing plant to run in a grid system.

---

[33] For detailed discussions of electricity in the conventional vision forecasts, see IEA 98, pp. 213–18; EIA 00, pp. 113–31; EC (POLES), Shared Analysis Project, vol. 8.
[34] See EIA 00, fig. 80, p. 114.
[35] See Felix Martin, 'Financial Reform', in *Energy and Development* (Washington DC: World Bank, 1999), pp. 6–12.

The conventional vision for electricity

Electricity demand is projected to grow faster than total energy demand in all regions and all three projections, but the rates of growth projected are below historic rates in many cases. The move towards 'high-tech' and tertiary industries involves increasing demand for electricity. In the conventional vision oil will still be used in increasing quantities, though with a diminishing share. Its use will not be confined to countries without cheap coal or natural gas resources: EIA 00 shows 10m b/d of oil going into power generation in 2020, half of which will be in developing countries and nearly 30% in industrial countries. Even in the United States a few oil power plants were still being planned in 1997.[36] Otherwise, the points of interest in the electricity sector projections are the growth of natural gas, the constraints expected to apply on nuclear development and the failure to achieve any significant increase in the proportion of other non-fossil-fuel resources. These trends are shown in Figure 2.9.

Figure 2.9: Change in share of fuel input for world electricity generation, 1995–2020

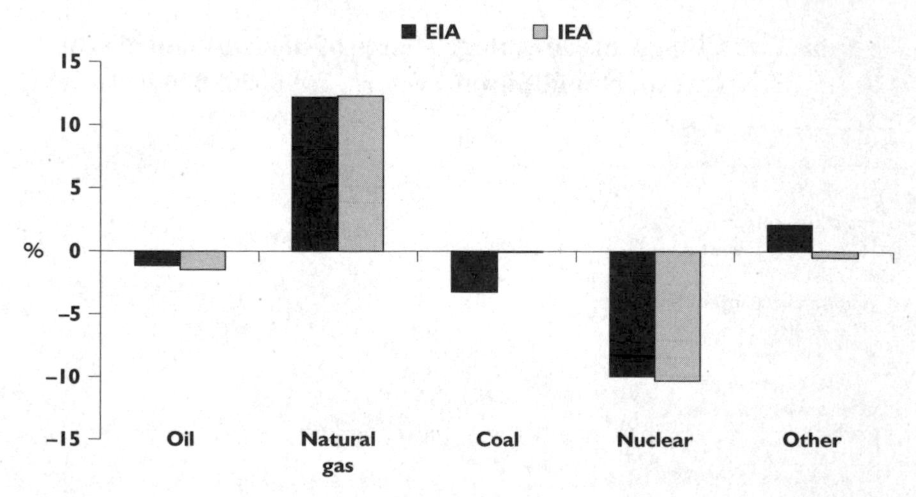

Note: Primary fuel equivalent basis.

[36] IEA 98, table 12.9, p. 214.

Other sectors

The conventional vision in EIA 00 sees oil demand outside transport and power more than doubling between 1995 and 2020 – a more rapid increase than for transport or for power.

Calculating the market outside transport and power as a residual probably exaggerates definition differences among projections. Nevertheless the conventional vision, though ragged on this point, is of oil losing share, despite an absolute growth in oil volumes. The large differences among the projections for gas demand in developing countries are apparent here.

Table 2.4: Oil use outside transport and power

Sector	1995 (mb/d)	2020 (mb/d)	Increase (%)
Oil (all sectors)	69.9	112.8	61
Transport	33.5	62.5	87
Power	6.6	10.3	56
Other	29.8	72.8	144

Source: EIA 00.

Table 2.5: Change in percentage shares of demand outside the transport and power sectors, 1995–2020

Fuel	EIA (*IEO-00*)	IEA (*WEO-98*)
Oil		
Industrial countries	–6	–2
Developing countries	–11	0
Gas		
Industrial countries	8	2
Developing countries	17	8
Coal		
Industrial countries	–2	1
Developing countries	–6	–8

Energy prices: difficult theory, poor data

Energy is an abstraction. In the 'new economy' of which oil is a part there are many fuels in chains of services and processes which deliver a variety of services to users, and the trend is for prices to be established by competition at every stage on each value chain and among different chains of value to consumers who have ever-expanding choices. Different fuels used in different processes deliver energy (usually heat or work) to consumers, and pollution to their and usually also other people's environments. The prices of fuels reflect the costs and/or benefits associated with the energy they carry. An energy price is also an abstraction, and across different fuels can only be an index number of fuel prices (which may or may not incorporate all their non-energy properties), weighted by the energy content of the fuels.

Econometricians look at the historical record of changes in fuel prices and quantities to distinguish several factors:

- *An income effect or 'rebound effect'*. Increasing (reducing) fuel prices reduces (increases) the purchasing power of consumers' income: higher incomes caused by lower prices will increase energy consumption; the consumers' allocation of the increased income to energy purchases may reduce as income rises. Thus income may be heading in a different direction from fuel prices: the innumerate management observation that the effect of fuel price changes 'wears off' when incomes are rising means simply that rising incomes have increased demand. Reducing the cost of using energy through 'win–win' efficiency measures causes a similar problem, the so-called 'rebound effect'. In both cases demand will be less than if the price or efficiency had not changed.
- *An efficiency or substitution effect.* An increase in fuel prices may cause consumers to spend more on new equipment, building materials and management operations, which will reduce the amount of fuel required to give the same energy result to the user.[37] The extent of the efficiency effect depends also on what happens to the price of the new equipment or building: if those prices rise in line with the fuel

[37] See Atkinson and Kehoe, 'Models of Energy Use'.

price, changes in the balance between fuel and capital or management will not occur. A new user technology (such as the development of the combined cycle gas turbine generator) may increase efficiency and thus greatly reduce the quantity of primary fuel needed to produce the required output (in this case electricity). If electricity prices had remained 'sticky', and the electricity and gas markets were not competitive, some of this advantage could have accrued to the gas supplier in the form of an increase in price – because the unit of gas produces more output of electricity, it would have a higher value. In reality, the development of competitive markets in both gas and electricity in the United States and Europe (now followed in Japan) has tended to ensure that the benefits of such technical advances accrue to the consumer through lower final prices. The same may apply in the case of improved efficiency in vehicles: the consumer's cost of motoring is reduced.

- *A fuel-switching effect.* Substitution of a cheaper fuel (possibly at some cost in changing equipment), may mean that the effect on energy demand of changes in the price of one fuel is different from what it would be if no switching were available – always assuming that the price of the other fuel does not change. The growth of oil demand in the decade up to 1973 had this characteristic. Not only was the imported oil often cheaper than domestic coal on an energy-equivalent basis, but the cost of transporting and handling oil (and dealing with its emissions) was less than that of coal. Again, competition among suppliers tended to ensure that the cost saving accrued to users rather than to the oil supplier: oil prices fell during this 'supply-push' expansion of the oil market.
- *The effect of the cost of constructing capital equipment or buildings which it would be costly to replace merely to save fuel.* Energy costs are usually a relatively small fraction of total operating costs of shops, offices and factories, except for a few industries such as aluminium smelting, steel, cement manufacture and paper-making, and commercial sectors such as food storage and distribution. Except for poor households, similar considerations apply in the domestic sector. (The exception here is important. Governments are reluctant to impose increases on fuel costs on poor consumers who cannot easily

avoid paying them.) A change in technology or fuel which would be justified in 'greenfield economics' of new equipment or buildings may not be justified when an existing plant can be used at variable cost, and may not be an option at all for businesses operating on narrow margins or households on low incomes. The need for capital purchases (house, car or manufacturing equipment) limits the efficiency option as long as the old capital lasts – that period varying with the age of the equipment and the particular relationship between capital and fuel costs.[38] This phenomenon leads to the frequently observed differences between price elasticities estimated from time series, which tend to be lower than those estimated by cross-country studies: the presumption is that the country observations capture the effect of sustained differences in price on the technology of use and substitution.

Sometimes everything does work in the same direction: the very large oil price increases of 1979–80 involved an absolute transfer of income from oil importers to exporters. The response after the first oil shock of 1973 tended to be to accommodate this by inflation and 'recycling' of the exporters' financial surpluses, but real incomes in the importing countries fell nevertheless, with an effect on demand. After the second oil shock monetary policy was set against inflation and there was a recession leading to a loss of output and income. This combined with the delayed 'efficiency' and switching responses to the first oil shock. The effect on oil demand in the mid-1980s was very large, leading (apart from economic damage to the importing economies[39]) to surpluses of capacity in the oil production and refining industries.

It is difficult to disentangle all these effects: there were so many fuel prices, some of them controlled by governments or by previous contracts for some of the time. Either the prices of efficient alternative plant, vehicles and building were not known, or the supply responded slowly so that price incentives were weakened. In the particular case of

---

[38] Ibid.

[39] See e.g. Knut Mork and Robert E. Hall, 'Energy Prices, Inflation and Recession', *Energy Journal*, vol. 1, no. 3 (1980).

electricity, price effects were dominated by the structure of the industry, the regulations applied to it, and the problems of over-capacity following the economic slowdown in the OECD area in the early 1980s.

Oil prices: no longer in the lead

It is conventional to view oil price projections as the centre of 'business as usual' scenarios for both energy producers and consumers. Debate has focused on the balance of demand for oil against the supposed availability of supply. Less attention has been given to major shifts of demand among fuels, and to the price relationships among fuels necessary to support such shifts. The problems of projecting the supply of oil are dealt with in the next chapter of this book (Chapter 3). An attempt to connect oil prices with future competition among fuels – the 'new economy' of oil – appears in Chapter 6. The need for such an attempt is shown by the difficulties of reconciling two features of the conventional vision which are described below.

In the conventional vision, the share of natural gas in primary energy demand will increase significantly in most countries and sectors (partly at the expense of oil), while gas prices will increase relative to other fuels, particularly oil. Current 'business as usual' scenarios now avoid the strongly increasing trend in oil prices which was formerly characteristic of many projections, and the long-term projections are lower than they were five years ago. IEA (*WEO*-98) projected an increase to over $30/bbl (1999$) by 2020, and in 1995 the same agency, in its 'capacity-constrained' case (there was no reference case), projected a price of $31/bbl (1999$) by 2005. The EIA 2020 reference projection of $22/bbl (1999$) is just below their projection for the previous year and significantly below the $28/bbl (1999$) projected in their 1995 *International Energy Outlook* as a reference case for 2010. The EC (POLES) reference is almost $28/bbl (1999$) in 2020, compared to a 1992 EC projection of $34/bbl by 2005.[40] Among the present

[40] European Commission, Directorate General for Energy, *A View to the Future: Energy in Europe* (Brussels, European Commission, September 1992), p. 25: assumption for average Community import price.

Figure 2.10: Oil price projections compared to past averages

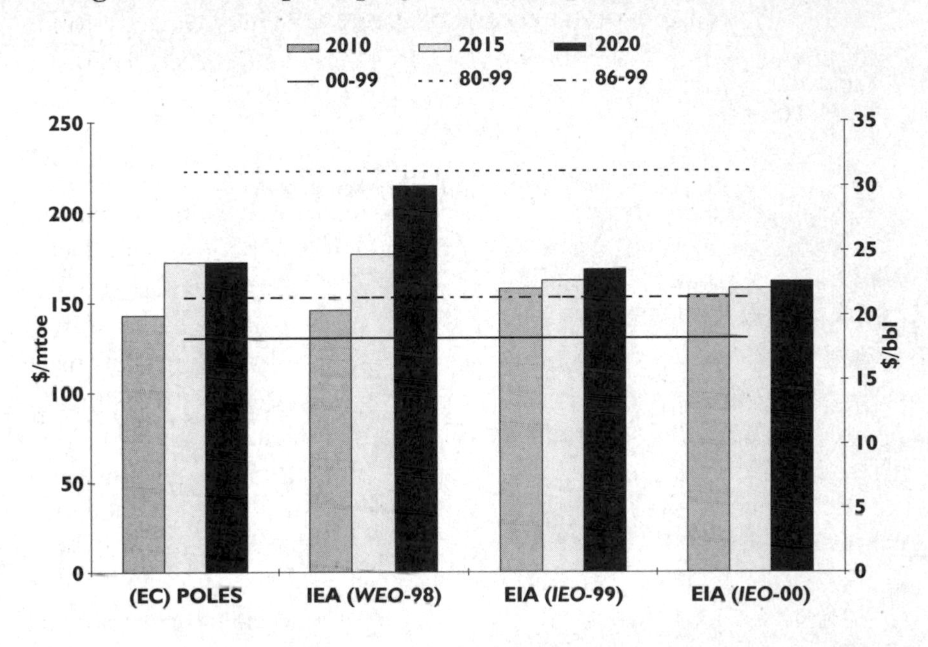

projections, only the IEA (*WEO*-98) projection approached the average of the period 1980–99.

Prices and market shares

It is difficult to avoid the evidence that changes in oil prices have in the past affected the share of oil in relation to other fuels in the energy mix, even at a global level.

Figure 2.11 shows the relationship between the oil price and the oil share of world energy consumption since 1965. The EIA (*IEO*-00) projections to 2020 are shown. The ten-year average oil price is used, partly to smooth short-term fluctuations and partly because of the need to demonstrate the idea of a lag in adaptation by consumers and by the suppliers of alternatives. Ten years is not an econometric estimate of the average lag in the effect of prices. At national and regional level, account would have to be taken of important details such as the availability of alternatives, and the past investment in capital and buildings

Figure 2.11: Fuel shares of world primary energy consumption and ten-year average oil price, 1965–2020

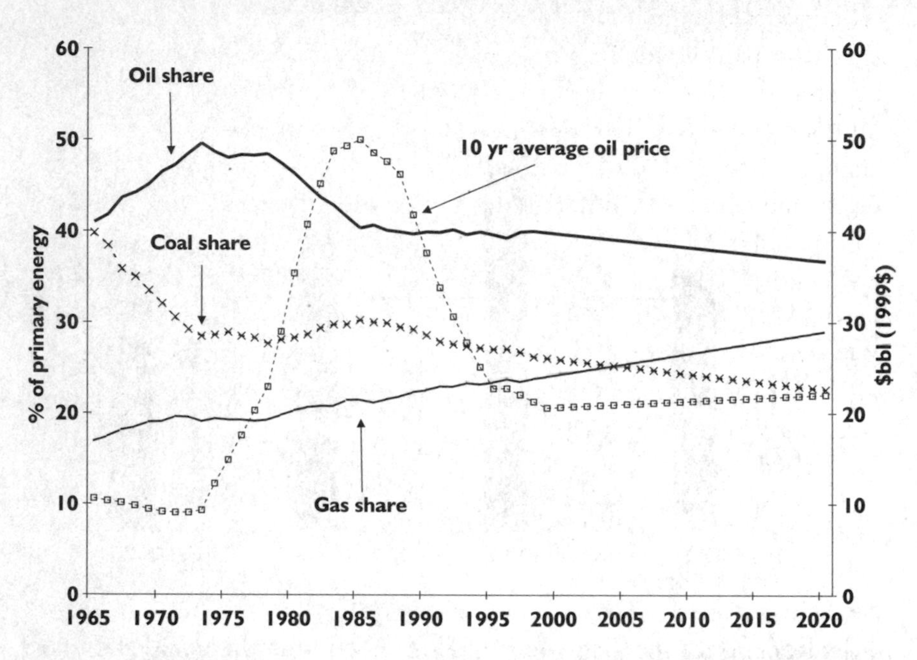

Source: 1965–98 data, BP Amoco *Statistical Review*, 1999; 1999 data, EIA *Annual Energy Review* 1999; projections interpolated to 2020 projection in EIA (*IEO*-00).

by users of particular fuels. These cannot sensibly be aggregated globally.

There appear to be four phases:

- *The first phase, to 1973,* showed a rapid capture of market share by oil. Oil prices had fallen slightly in real terms for over a decade. It was, in effect, a 'superior fuel' relative to the others. Oil demand grew at 8% annually while energy demand grew at 5% annually and its share of world energy consumption grew by 10% in the decade to 1973. The trend was unsustainable for many reasons.[41]

[41] For a discussion of the interaction between price and fuel share, see J. E. Hartshorn, *Oil Trade: Politics and Prospects* (Cambridge: Cambridge University Press, 1993), pp. 2–8, 31–7.

- *The second phase, 1973–85,* was one of dramatic structural change in the oil market.[42] The period began with most international trade in oil conducted within and among about a dozen international companies who paid governments of exporting countries a tax based on a negotiated reference price. The period ended with most international oil trade being priced on the spot market with formal commodity exchanges for oil in New York and London. The surges in prices were large, and occurred over a shorter period than the lead times needed for practical changes in demand or competing supplies. World energy demand fell by 3% between 1979 and 1983. Over the whole period from 1973 to 1985 oil's share of energy demand fell from 50% to 40%. Major industrial countries' energy demand fell and oil's share fell further: the coal share stabilized and the gas share continued to rise gradually.
- *The third period, 1986 to date,* has included periods of volatility around 1990–91 (provoked by the Gulf War) and in 1998–99 (in response to a combination of events including the effect on demand of the Asian financial crisis of 1997–98, and the action of OPEC in first increasing, and then reducing, production quotas against the swings in demand). The oil share of world energy consumption more or less stabilized, while that of gas continued to rise and that of coal resumed its earlier fall. The total fossil fuel share fell from 93% to 90%.
- *The fourth phase, to 2020,* is covered in the figure by projections from the EIA (*IEO*-00). Other projections are compared below. On this projection, oil would lose 3% of market share between 1998 and 2020; coal's share would continue its relative decline, also losing over three points of market share over this period (but while coal use increased by 50%); gas would gain 5% in market share, while the fossil fuels together would lose nearly 2%.

How do these shifts compare with the projected shifts in relative prices?

[42] See John Evans and Gavin H. Brown, *Opec and the World Energy Market* (Huntingdon: Longmans, 1986; 2nd edn 1990).

Gas and coal prices

The EIA (*IEO*-00) reference oil price projection for 2020 is only 10% above the 1986–99 average and over 30% below the 1980–99 average (see Figure 2.10). Big future movements in the oil share, if driven by prices, would have to be driven by movements in the prices of competing fuels. However, in the EIA projection (*IEO*-00) the ratio of the wellhead gas price to the projected crude oil price rises slightly from its average for 1986–99. The mine-mouth coal price falls in the EIA reference projection, so that mine-mouth US coal becomes relatively cheaper than oil: one-sixth of the oil price in 2020 compared to about one-quarter in 1986–99 (the fall is not as strong at the point of use, because of the higher proportion of coal projected to be produced in the western United States). The price movements and share movements appear to work in contradictory directions in the EIA projection, but not strongly so: other local factors – for example, the costs of removing sulphur from coal and residual oil to compete with gas – could explain the inconsistency.

In the other forecasts, the contradiction between the projected relative prices of oil and gas and movements in relative market shares is more difficult to avoid. Figure 2.12 compares the 1986–98 average and the projections for 2020. In the EC (POLES) projection the quoted gas price increases to almost equal the crude oil price, from half the crude oil price in Europe and the United States during 1986–98. For Asian LNG, the increase is from two-thirds of the 1986–98 average. The IEA (*WEO*) relative increases in gas price are less dramatic, but still (except for the United States) large enough to raise the question of consistency with the increases in gas market share.

For coal, the EIA (*IEO*-00) projects the ratio of coal mine-mouth prices falling to a quarter of gas wellhead prices: on a delivered basis the fall would be to a third.[43] The EC (POLES) projects a fall (from three-quarters to one-half) in the ratio between coal and gas import prices for Europe, and a similar fall in Asia. IEA (*WEO*-98) projects a rise in the ratio of coal to gas import prices in both Europe and Asia –

[43] *US Energy Outlook 2000*, table A3.

Figure 2.12: Gas price projections for 2020 compared

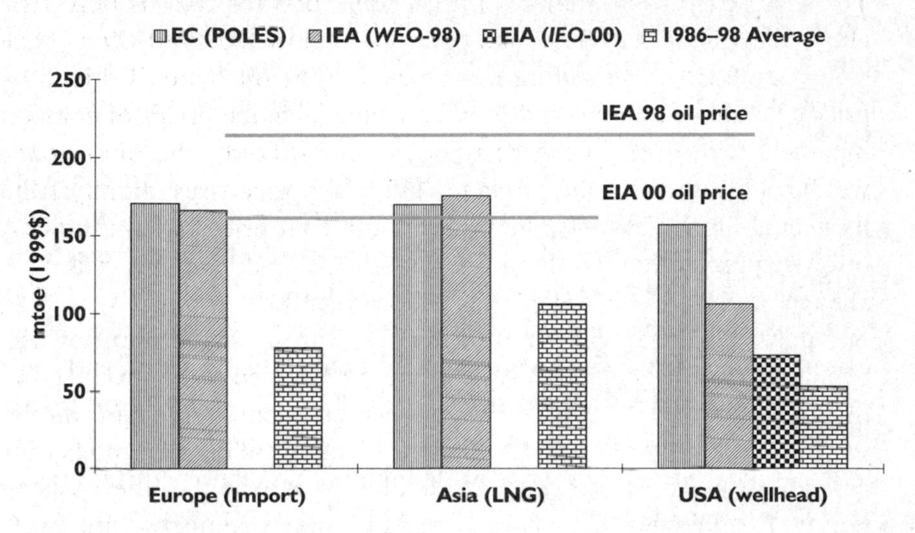

Note: Not all sources provide projections for the same (or all) prices.

consistent with its projections for a major shift from coal to gas for power generation in Europe and the OECD Pacific (but not in the rest of the world, where coal supplies are mainly indigenous).

Conclusions: the conventional vision and its problems

The question posed at the beginning of this chapter was whether the energy trends in the 'reference' case projections of the three main public agencies contain uncertainties, discontinuities and contradictions that may give rise to changes in the predicted trends. The conclusion of the analysis is that most uncertainties, and the greatest apparent inconsistencies, concern the role of natural gas in the energy mix and, connected with that, the price of natural gas relative to other fuels.

There are also differences between the projections concerning regional economic growth, with the EIA (*IEO*-00) projecting higher growth for the United States than the other two, earlier, non-US projections. The US projection is also more conservative regarding future energy intensity. Getting the regional balance wrong is more directly important for gas and coal, where different regions and countries have different local resources, and transport costs matter more than for oil. The demand for oil will be affected by regional differences in economic growth through the different degree of competition from other fuels in each region and though the different weight of transport fuel demand in energy consumption.

The conventional vision of future oil prices is of modest growth: not much above the 1986–99 average, and below the 1980–99 average, with the oldest projection (IEA *WEO*-98) having the highest projection for 2020 and the newest (EIA *IEO*-00) the lowest. These projections involve assumptions about the growth of oil supply which are more optimistic than those current in conventional thinking as recently as five years ago, and which are discussed in Chapter 3. These oil price projections are broadly consistent with the view that the oil share of world primary energy use will fall, though the projections differ about the degree of decline, which countries will be most affected, and which sectors will lose most. In the EIA (*IEO*-00) projection for 2020 the oil share of the North American primary energy market actually increases slightly, probably due to the greater weight of transportation in the primary energy mix. Strong growth in the demand for oil for transport is projected by all three agencies for all countries. Question marks over the sustainability of such growth are discussed in Chapter 4; they arise not from the *availability* of oil for transport, but from its acceptability and the rising competition from alternative transport technologies.

The EIA and EC (POLES) projections show a fall in coal prices, significantly in the United States, less so elsewhere, while the EIA (*WEO*-98) projects a rise in prices for coal imported to Europe and Asia. All projections assume a fall in the coal share of the primary fuel mix. Since all projections also assume electricity demand growing faster than energy demand in all regions, this is mainly explained by a fall in the coal share of energy use outside the electricity sector. Coal's share of electricity supply in North America, the Pacific OECD and many developing countries gives a stability to its overall share of primary energy consumption in many countries. Europe is the main exception, presumably because of the effect of current and expected environmental policies on investment in coal generating plants and the effect of regulation in providing an opening for new, independent gas-fired power generators.

For natural gas, the conventional vision is more complex. Natural gas is seen as gaining market share from oil in many regions and sectors, despite a projected increase in its price relative to the prices of other fuels. The least contradictory projection on this point is EIA (*IEO*-00), where the relationship between gas and oil prices does not change, but the price of coal falls relative to both. The projections deal in producer or import prices, rather than prices to final consumers. It is possible that the costs of gas transportation and distribution, and the profit margins on these activities, could fall so far as to negate the effect of the change in price relationships which is projected at the producer and import level. Such a development would be consistent with the experience in the United States and Britain, and now beginning in the EU and Japan, as a result of restructuring the regulation of the electricity and gas markets in order to promote competition at every stage of the supply chain. Similar developments in many developing countries could, however, have an opposite effect where those countries have previously maintained price controls and subsidies which have kept prices down. However, in these countries, lack of availability due to lack of investment under price restraints may also have restrained demand so that economic reform which raises prices may also stimulate investment to meet demand previously frustrated by lack of supply. The position of gas in the Russian Federation and Ukraine, where subsidies take the form of non-payment, is a special case.

The question which arises from the projections for gas may be summed up in the following terms: If gas prices really will rise relative to other fuels at the point of competition, will gas make the predicted rapid gains in market share? The answer may depend on the value attached to non-energy qualities of gas or on changes in market structure. On the other hand, if competition among producers and in gas transportation to new markets means that price of gas at the producer and import level does not rise so strongly relative to other fuels, are the demand and market share growth projections for gas perhaps significantly understated? These contradictory possibilities are discussed in Chapter 5.

Chapter 3

Oil supply

Introduction: the future of oil supply

The twentieth century progressed hand in hand with increased usage of oil, in times of both war and peace, and can justifiably be described as the Oil Era. Oil's role as a fuel in the twenty-first century is less clear. In most cities in the industrialized world, ground-level air pollution caused by burning oil is falling, despite its increasing use, as products are cleaned of pollutants and more thorough combustion systems are introduced to boilers and vehicle engines. However, a new threat is emerging, namely the global warming effect of carbon dioxide released to the atmosphere when fossil fuels are burnt. Concerns about climate change are beginning to influence policy relating to energy generally, and transportation in particular. Questions are being asked again about how much oil remains to be produced. There is a paradox here: the less oil remains, the less of a threat its extraction is to the climate, but the greater the probability that alternative fuels will either be more environmentally damaging (coal) or more expensive (nuclear), or will require massive investment in infrastructure (gas for developing countries).

The current consensus is that the world's conventional oil production will peak during the period 2010–20, and inevitably decline thereafter, although some see the peak before this date, and others as far away as 2050.[1] The key factors influencing supply-side economics are:

- underlying geological knowledge;
- how far technology will increase the recovery of discovered oil, and aid discovery of further reserves;

---

[1] In line with the IEA's definition, oil is considered conventional if it is produced from underground hydrocarbon reservoirs by means of production wells and/or it does not need additional processing to produce synthetic crude.

- the extent to which technology and managerial change will continue to reduce the costs associated with exploration, production and transportation;
- the abundance of unconventional oil (synthetic oil, oil sands and oil shale), which already forms part of world production;
- the policy conditions and trends in taxation that increase or reduce the demand for oil, particularly with relation to policies implemented to mitigate the effects of climate change;
- constraints on accessing oil that may lie in environmentally or politically sensitive areas; and
- how the structure of the industry affects investment and the development of supply (in most oil-exporting countries the industry is under government ownership or control, yet organizational and technical advances tend to originate in private sector companies).

The reserves debate: half full or half empty?

Sensible policy depends on an understanding of remaining reserves, yet this number cannot be known for sure. Indeed, the degree of uncertainty inherent in any estimation of oil reserves is compounded by the poor quality of data on which all such estimates are based. Also, there can be little doubt that reserve estimates, given their very nature, have been, and will continue to be, subject to the political manipulation of governments and to the commercial considerations of companies. The clearest example of this was the restatement of OPEC reserves in the late 1980s. In 1987 alone, world proved reserves increased from 700 billion barrels (Gb) to 900 Gb as a result of substantial revisions by Iran, Iraq, Abu Dhabi and Venezuela. Saudi Arabia followed suit in 1989, adding another 85 Gb by itself. Was this a correction by governments of previous figures that had been understated by the oil companies who held concessions prior to the 1976–79 'participation'? Or were governments leaning towards optimism in declaring reserves for the purpose of negotiating OPEC quotas?

The problem is partly definitional. 'Remaining reserves' means that quantity of the in-place resource that can be recovered economically. In the United States the definition is strict, and is based on drilled reservoirs and current prices. Elsewhere the owners or estimators have more discretion in what they declare. 'Proved reserves' – the definition of reserves used in the United States – are those in discovered fields only, estimated with a high degree of certainty. This typically does not allow for additional reserves recovered through secondary or tertiary recovery techniques unless such a mechanism has been shown to work locally. As a general rule, this method makes for good, or p50, estimates of reserves in fields where production is already in decline, but tends towards overly conservative (p90) estimates for fields yet to reach peak production.[2] The standard also requires an economic test: prudent accounting dictates that only those quantities that can be produced at today's price and costs can be included. Thus, when prices fell in 1998, so did estimates of proved reserves. They rose again in 1999 when prices increased. Indeed, any reserves estimate depends on assumptions made regarding future price and costs.

This definition is very conservative and not universally applied. Standards vary across countries, with probable and possible reserves often added to those that are proved. Healthy scepticism is appropriate even when estimates come attached with a probability estimate (p50, p90, etc.) Mean values stated by the United States Geological Survey (USGS) in its *World Petroleum Assessment 2000* essentially confirm the central value of 2,300 Gb used in the IEA's 1998 *World Energy Outlook*. In this survey the USGS also quantifies the future reserve growth from conventional reserves, and estimate this to be almost 700 Gb, thus increasing total ultimate reserves to 3,021 Gb. If all natural gas liquids (NGLs) are included this number increases to 3,345 Gb.

A useful concept is to think of all the oil in the world as an iceberg whose visible top contains the volume of prime *reserves* that are of high quality and easy to extract. The spotlight of most policy analysis is focused on these proved reserves. The question 'How many years until the oil runs out?' is usually answered by looking at this volume of reserves. On this basis the world's oil reserves could supply needs for

---

[2] By definition, a p50 estimate has an equal (50%) probability of being either too high or too low. A p90 estimate has a 90% chance of being too low.

Figure 3.1 The oil iceberg

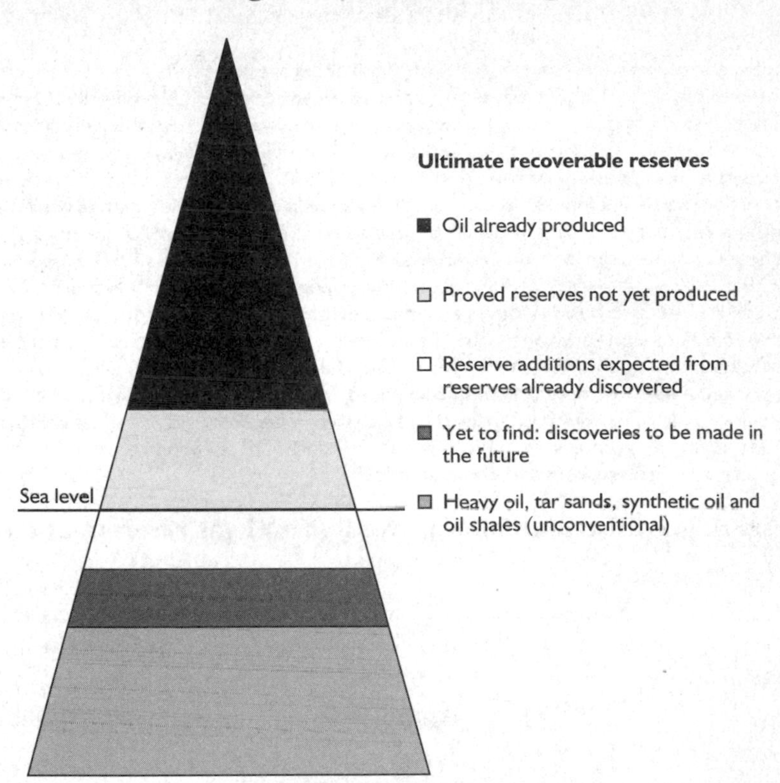

over 40 years, or about 80 years of current oil consumption for transport. Below 'sea level' there is a larger oil *resource* that is of lesser quality or is more difficult or expensive to extract. The iceberg image has the advantage of suggesting what history has shown: that reserves are not fixed, but move up (and occasionally down) over time. The iceberg is cut at different heights, depending on economics and technology. What stands out above the cut are only the proved reserves. Below the cut is the resource from which reserves are continuously being replenished.

The world is not about to 'run out' of oil. We are perhaps approaching the mid-point of consumption of what are sometimes called *conventional* oil reserves, as defined by the economics and technology we now have. The rate of consumption of this visible part of the iceberg will

Box 3.1: Gas reserves

Natural gas reserves are similar in scale to those of oil but are more widely distributed across all regions (see Figure 3.2). Proved gas reserves are 62 times current annual production. Growing demand and technological advances will allow conversion of more conventional gas resources into additional reserves that are economic to produce. Where conventional gas reserves have been, and remain, low relative to demand (e.g. in the United States, where the proved reserves/ production ratio is nine years), tax incentives have stimulated the development of unconventional gas, such as coal-bed methane, which now accounts for about 7% of US reserves. A gas 'iceberg' would look similar to the oil picture (see Figure 3.1), provided a slightly smaller fraction were shown as already produced, and two types of unconventional resources were excluded. The first exclusion would be methane hydrate or clathrate resources, since their future commerciality remains in doubt. These resources are simply enormous. Estimates suggest they are an order of magnitude greater than the total of all conventional gas reserves. They could possibly be developed in the longer term, but their diffusion, or lack of density, poses a formidable technological and economic challenge. Secondly, gas resources held in solution in aquifers – geopressured gas – which are considered to be even larger than hydrates, are also excluded on grounds of uncommerciality.

Figure 3.2: Distribution of proved oil and gas reserves at end 1999

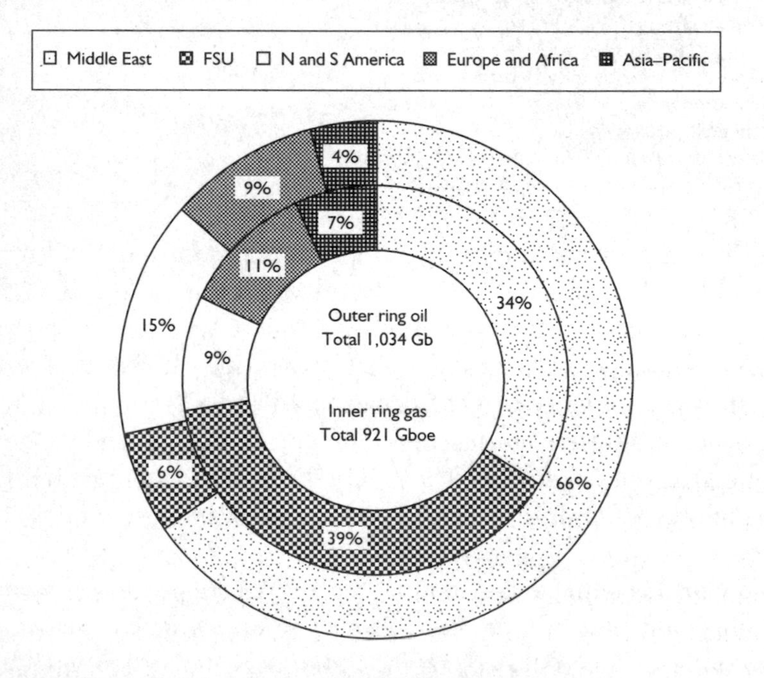

Source: BP Amoco *Statistical Review of World Energy 2000.*

Box 3.1: Continued

The high transport costs of gas, relative to other fuels, have meant that in most regions high and growing gas shares in the energy mix have been associated with the presence of an indigenous gas supply and the absence of alternatives such as low-cost domestic coal or nuclear power. Even where such alternatives exist, as in the United States, imports are expected to increase (through new pipelines from Canada and re-opened liquefied natural gas (LNG) terminals to receive shipping from Central and South America and West Africa) as technological developments continue to bring down both LNG and pipeline costs. BP Amoco, for example, has stated that the costs of the Trinidad LNG project will permit supply into the United States at prices that are competitive with pipeline supply (*c*.$2/mm Btu).

Europe's main natural gas sources (the UK, Norway, the Netherlands, Algeria and Russia) have sufficient gas reserves to support growing European demand in the next decade. There is a well-developed transport infrastructure and currently an over-supply of both gas and transit capacity. However, the extent to which the present surplus can carry Europe through following decades is uncertain. It is possible that additional supply sources and infrastructure corridors will be needed from further afield, such as the Caucasus, Central Asia and the Middle East.

The development of an international gas pipeline infrastructure in South America in the 1990s demonstrates the potential for the growing use of gas in the developing world. The availability of natural gas makes developments possible where there is also a will to solve transit problems and other issues of viability, security and international interdependence. Where distances are too long to justify cross-border pipelines, LNG will be supplied in ships. The LNG trade has been growing at a rate of 7% annually, and can be expected to continue in this trend.

New technologies are under development – for example, the use of fuel cells – which have the potential greatly to expand markets for gas. Technology is also allowing pipelines to be laid in deeper water, bringing the prospect of gas supply to yet more populations. Furthermore, processes for converting gas to liquids offer opportunities to market over 7 trn cu metres of natural gas in over 2,000 fields which are currently isolated from traditional gas markets. That such developments are already in train is not in doubt. What is uncertain is the extent of their impact on gas and oil markets over the next 20 years.

slow as pressure in the producing oil fields declines, the oil flows more slowly, and it becomes more expensive to increase the fraction of the reserves (currently between 30% and 40% worldwide on average) that is actually extracted from reservoirs. This does not mean total oil production itself will decline: the focus of reserves will shift down to the hidden resources. Increasing amounts of *unconventional* oil will be produced from heavy oil and tar sands. More synthetic oil will be produced from coal and from gas reserves located far from markets. Perhaps the price will rise sufficiently to allow development of parts of the vast shale oil base.

Since reserves are like stock in a warehouse, continuously being replenished from resources, it is wrong to think of reserve/production ratios as indicators of when oil will 'run out'. While these ratios are useful *comparative* measures, they have little meaning in an *absolute* sense. Oil production is more likely to peak when cheaper alternatives become available than to be constrained by its own finiteness. The shape of the oil production curve is thus controlled as much by competition from other energy sources as by its own availability.[3]

The pessimistic view

Colin Campbell and others of his school (referred to as 'pessimists' in the IEA's 1998 *World Energy Outlook*) estimate the world's ultimate (i.e. including past production) *conventional* oil reserves to be around 1,800 Gb.[4] This figure could be stretched to 2,000 Gb if higher former Soviet Union (FSU) and Middle East figures were accepted. The views of the pessimists are grounded in the accepted belief that oil resources are non-renewable and finite. Reserves are seen essentially as static. The peak period for making new discoveries (the 1960s) is long past, and estimates of reserves yet to be found are low. Furthermore, the rate at which discoveries are currently being made (<15 Gb annually during the 1990s) is too slow to affect significantly the time when production will start to decline. It is argued that improvements in recovery techniques merely bring forward the time in which reserves are exhausted.

Key to the pessimists' view is that production will follow a variant of the Hubbert curve. A geologist with Shell Oil, M. King Hubbert, believed that resources are finite, and that they are depleted over a period determined by the quantity of reserves and the rate of production. Once

[3] See Peter J. McCabe, 'Energy Resources – Cornucopia or Empty Barrel?', *AAPG Bulletin*, vol. 82, no. 11 (November 1998), pp. 2110–34.

[4] For a summary of the 'pessimist' viewpoint see Colin J. Campbell and Jean H. Laherrère, 'The End of Cheap Oil', *Scientific American*, March 1998. More detail is to be found in Campbell's *The Coming Oil Crisis* (Brentwood, Essex: Multi-Science Publishing and Petroconsultants, 1997). It is worth pointing out that these modern-day proponents of scarcity make the same arguments as used in the 1970s by Warman and others.

around half of the reserves have been produced, declining production is inevitable. On the basis of this belief, in 1956 he predicted that onshore US production would peak between 1965 and 1970.[5] In an individual reservoir, there are technical reasons why production might follow this curve and eventually become uneconomic. Note that this implies an economic, rather than a physical limit to reservoir production. Also, as Morris Adelman has pointed out, this curve is typical of any product, with demand growing and then falling as a result of obsolescence and competition.

Hubbert's idealized production curve is bell-shaped, though he recognized that actual production plots may differ greatly from this form. Production in the United States illustrates this. In the Lower 48 United States, the combination of falling new discoveries and most known reservoirs having passed the halfway point in their recovery meant that, as Hubbert correctly predicted, production began to decline in 1971. But the discovery and development of a new province (Alaska) and the technology permitting the development of deepwater reserves in the Gulf of Mexico temporarily reversed the decline. The United States is a paradigm case: Alaska is now also in decline, and large areas of potential are closed to new exploration for environmental reasons. The world is following the United States as mature provinces (e.g. the North Sea) face decline and not enough new discoveries are made. As there are only a finite number of sedimentary basins in the world, eventually all likely potential areas will have been explored and developed. There is already a long list of basins where the industry held high hopes of discovering large new sources of supply, but where dreams have disappointed: a few recent examples include China's Offshore and Tarim basins, Vietnam, the Atlantic Margin and the Falklands.

With cumulative consumption by the year 2000 estimated to be close to 900 Gb, and estimated ultimate reserves of 1,800 Gb, the pessimists' logic has it that the peak of conventional oil production will be reached between 2000 and 2010. Oil supply will stall and the world will have to cope with a future of fairly rapid decline in supply.

[5] See M. King Hubbert, 'Energy from Fossil Fuels', *Science*, 4 Feb. 1949; also the website <www.oilcrisis.com>.

The IEA view

The IEA's 1998 supply outlook may be characterized by the term 'relative abundance'. Its latest assessment of reserves takes 1,800 Gb as a starting point, but recognizes that technological progress is likely to increase this to between 2,000 and 3,000 Gb, with a most likely figure of 2,300 Gb.

On the basis of these numbers the IEA does 'not foresee any shortage of liquid fuels before 2020, as reserves of unconventional oil are ample, should the production of conventional oil turn down'.[6]

Any decline would coincide with that of the OPEC producers of the Middle East who currently hold most of the world's stock of reserves (see Figure 3.2). In the IEA's view, Middle East production is not projected to rise above 48m barrels per day (b/d), although this itself depends on the investment required to double current production of 20m b/d to 40m b/d by 2010. By 2014, the year of peak production in this central case,[7] Middle East OPEC's share of world production will be approaching 50%. The comparable figure in 1973, at the time of the first oil price shock, was 35%. In 1999 it amounted to just 28%. The period during which the Middle East would hold increased market power is expected to be limited, since this share is then expected to decline as unconventional oil becomes increasingly needed to replace and augment declining conventional supply (see Figure 3.3). Modelling work performed by the IEPE (the Institut d'Economie et de Politique de l'Energie in Grenoble, France) confirms that unconventional oil emerging in regions outside the Middle East could restrict the region's share of world production to around 40% after 2020. Around 20m b/d of required production from unconventional sources will need to come from developments yet to be identified.[8] A growing supply of oil thus depends on developments of unconventional reserves.

[6] IEA, *World Energy Outlook 1998*, p. 101.

[7] In the reserve sensitivity cases (2,000 Gb and 3,000 Gb), peak production occurs in 2010 and 2020 respectively.

[8] IEA, *World Energy Outlook 1998*, p. 103. The existence of unconventional oil is not in doubt. Which particular accumulations attract investment and development remains to be determined.

Figure 3.3: World oil production forecasts

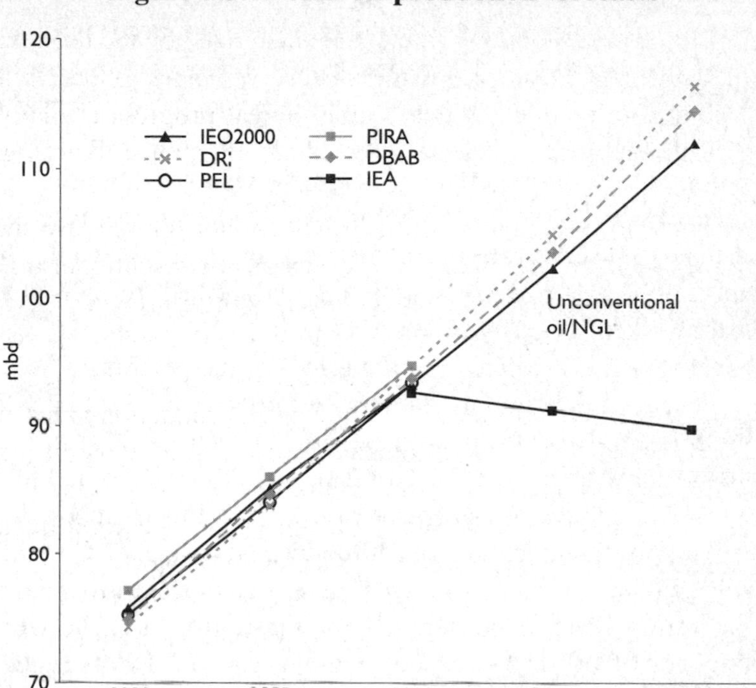

Abbreviations: IEO2000: *International Energy Outlook 2000*; DRI: DRI McGraw-Hill, a unit of Standard and Poors; PEL: Petroleum Economics Ltd; PIRA: Petroleum Industry Research Associates; DBAB: Deutsche Bank Alex Brown; IEA: International Energy Agency.

Source: IEA, *International Energy Outlook 2000.*

The optimistic view

To Peter Odell, one of the leading optimists, 3,000 Gb is a central estimate of conventional oil reserves derived from the trend of reserves estimates since the 1950s. His published work on this subject dates back to the 1970s, to the era of *Limits to Growth*, when he was one of the few to stand against the conventional thinking of the time that said we were rapidly exhausting fossil fuel supplies.[9] His view, based on

[9] P. R. Odell and K. E. Rosing, *The Future of Oil* (London: Kogan Page/Nichols, 1980; 2nd edn 1983).

these reserves, and long-term annual growth in energy use of 2%, is that the peak of conventional oil production will not occur until around 2030.[10] If just 3,000 Gb of the vast known unconventional resource base – oil shales, tar sands and other hydrocarbons – were converted to reserves, the real peak would be delayed until around 2060. While he does not include them in any of his figures, Odell also points out the existence of a large body of geological science in the FSU which claims that oil reserves are inorganic in origin and renewable. In the West this notion that reservoirs are being replenished from deep within the Earth is taken seriously by only a few.[11]

Michael Lynch has pointed out the errors of the pessimists' past predictions. His own assertion in 1989 that oil prices would actually fall in the 1990s as non-OPEC supply grew – called 'heretical' at the time – fits what actually happened.[12] As a result, pessimists have had to defer their estimates of the date of peak production. Furthermore, a 1997 study of supply functions in over 40 countries has shown that there is as much evidence of supply expansion as of contraction, with findings providing little or no support for the notion that the world is 'running out of oil'.[13] Lynch has argued that the Hubbert method fails because it takes recoverable reserves as fixed, whereas in fact the dynamics are rather more complex: of prime importance is the continuity of the mix of knowledge, technology and investment that sustains the process of exploration and production sufficiently to meet short- and medium-term demand expectations. Reserves depend on the interaction of this process, government policies and, finally, the price people are willing to pay for oil products. Since we cannot

[10] P. R. Odell, 'Fossil Fuel Resources in the 21st Century', *FT Energy Report*, May 1999. A shorter and updated version of this may be found in *Energy Exploration and Exploitation*, vol. 18, no. 2 (2000).

[11] In Odell's view there are many more who simply refuse to study the evidence – or even listen to the arguments – for this modern theory. He likens this situation to the time when the theory of continental drift was first hypothesized.

[12] Economist Intelligence Unit, *Oil Prices to 2000: The Economics of the Oil Market* (London: EIU, May 1989).

[13] G. C. Watkins and Shane Streifel, *World Crude Oil Resources: Evidence from Estimating Supply Functions for 41 Countries*, World Bank Policy Research Working Paper 1756 (Washington DC: World Bank, April 1997).

know future technology or prices, we cannot quantify future reserves. This should not be a concern, since it is these processes that are important. Ultimately, as Adelman has commented, 'oil resources are unknown, unknowable and unimportant.'[14]

Underlying issues and uncertainties

Price and the economics of supply

Implicit in any estimate of reserves is an assumption about the future price of oil. The episodic history of prices, and what we can learn from price movements in recent years and glean about future prices, is the subject of Chapter 6. It is sufficient to say at this point that oil is not a typical commodity, in that its price is not governed by the cost of marginal supply. There are orders of magnitude difference between the operating costs of large, low-cost fields (typically around $2/bbl), the tag on an average-priced barrel of crude ($20/bbl) and the price of a gallon of petrol at the pumps, which can, with taxes, exceed $200/bbl. There is thus a political as well as a market dimension to price setting, involving both producing- and consuming-country governments.

A large part of world oil supply is controlled by the governments of oil-producing countries whose economies and public finances are dependent on oil export revenues. This defines qualification for OPEC membership. The price fall and rise of 1998–99 shows that some non-OPEC oil producers, principally Mexico and Norway, are also prepared to constrain production in the face of very low crude oil prices. In setting targets for crude oil prices, producers must also consider an appropriate ceiling. The upper bounds of the price of oil over the medium and long term should reflect the cost of producing alternative supplies of oil, the price of competing fuels, and the cost of new technologies which threaten to introduce substitutes for oil, or to reduce energy consumption.

---

[14] 'The total mineral in the earth is an irrelevant non-binding constraint. If expected finding-development costs exceed the expected net revenues, investment dries up, and the industry disappears. Whatever is left in the ground is unknown, probably unknowable, but surely unimportant: a geological fact of no economic interest.' Morris A. Adelman, *The Economics of Petroleum Supply* (Cambridge, Mass.: MIT Press, 1993).

Box 3.2: Technological progress

Undiscovered reserves

Scientific advances in seismic techniques, understanding of depositional settings, basin modelling, sequence stratigraphy and play fairway analysis – together with the integration of these disciplines- are helping to locate more oil at lower risk and with lower cost, even in the fold belts and deep offshore.

The USGS's latest estimate for undiscovered oil is 724 Gb. A glance at Figure 3.4 shows that the prospect for further major discoveries is likely to be limited. There appears to be greater scope for growth in increasing recovery from existing discoveries.

Recovery factors (RFs)

Cheap and powerful computers allow us to 'see' more clearly where oil in discovered reservoirs is trapped and recover more of it. The development of drilling technology, via horizontal, multilateral and yet more complex wells, aids the productivity and cost-effectiveness of exploitation. The record for extended reach drilling has recently been stretched to over 11 km. RFs of unconventional heavy oil are also improving, with steam injection offering the prospect of 20% recovery.

The 1998 *World Energy Outlook* states that historic RFs have increased by 1% annually on the UK continental shelf. If this feature – which is analogous to productivity gains in other industries – is universal, a typical primary RF of, say, 30% would increase over a 25-year period by 0.3% annually to reach 38%, so adding 500 Gb to reserves. Work by Jean Laherrère presented to the 1997 IEA Oil Reserves Conference comparing RFs in giant fields with earlier studies also suggests that RFs have improved. Extrapolated, this suggests the average giant field's RF may exceed 50% by 2020. Even then, the remaining 50% of these fields' resources would remain in place, always subject to capture by advancing human ingenuity and technological progress. Already there are estimates that some North Sea field RFs will rise above 60%.

Costs

Technological progress and innovation in exploration, drilling and production, along with the creation of alliances between developers and contractors, have led to very significant reductions in costs and lead times as well as enhanced production from mature areas. Cost reductions come from advances made in other industries (e.g. computing), from economies of scale, from the learning curve effect and through technological breakthroughs. Comparison of the costs and timescales of similar size developments offshore the UK in the 1970s and the 1990s illustrates the extent of changes to date. The time taken from development start to production of first oil has typically halved, from four to two years, and costs have fallen by over 75%, or around 3% annually in nominal terms. Given continued advances in technology and the likelihood of continued pressure to reduce prices, it seems reasonable to expect further cost reductions. How much further these reductions can go is not clear. Some companies are aiming to reduce overall costs to $7/bbl by 2001. In a commodity business costs are a key indicator of performance, and competition among companies, together with adoption of industry best practice, will ensure there is always pressure to reduce them further.

The economics of oil production are also complicated by taxes. In countries where private sector companies produce oil, they are subject to tax and licensing terms set by government. As a general rule, countries where exploration risk is low take a high proportion of economic rent (through royalties, taxes, etc.), while countries where exploration risk is high must offer a larger share of rent to companies to encourage investment. Countries with high prospectivity that invite private capital can exact onerous terms from an industry keen to acquire reserves; an example of this is the Venezuelan *apertura*, the opening up of the industry to private capital during the 1990s. Rates of tax vary, but are generally much higher than apply in other industries. The average state take worldwide is around 65% of upstream profits, though it can be as low as 25% (Ireland) or exceed 90% in some countries (e.g. Malaysia and Venezuela). 'Progressive' fiscal regimes (where the state take is based explicitly on a share of the economic rent) are considered to be more equitable than 'regressive' regimes where more of the take is fixed (e.g. through royalties or bonuses) regardless of levels of profitability. There is a trend towards the more progressive type of regime.

The way in which government sets and adjusts the level of rent it demands from production activities influences the economic attractiveness of the activity and affects the pace of exploration and development. Economies of scale associated with the development of large fields generally offer large rents. Despite the fact that satellite fields can make use of infrastructure already in place, as oil provinces become mature and field sizes smaller, terms need to be ameliorated to encourage continued activity. In the UK offshore, for example, the taxation of new field developments has now fallen to the same level as any other business, with only corporation tax levied.[15] Other countries can be expected to adopt similar measures in the future to arrest prospective production declines. As tax 'takes' are reduced, the recovery of additional reserves will be increased.

To put production taxes in context, it should be stated that fiscal terms affect supply only marginally. They can be used to help depletion

[15] The fiscal impact on the economics of UK North Sea production has been the subject of study in the Department of Economics at Aberdeen University since 1973. The results of these studies have been published by Professor Alexander Kemp and others in a series of occasional papers (71 to May 1999).

policy, if this is a consideration for government. In this regard the contrast between the UK and Norway, within the same hydrocarbon province, is striking. They also have an influence on whether private investment is directed to one country or another, and step changes in terms can significantly influence activity level within a particular country. However, they do not greatly affect the general level of activity, which is more influenced by overall price levels and exploration attractiveness.

The future price assumed in the IEA's 1998 *World Energy Outlook* is $17/bbl (1990$), equivalent to over $21/bbl in 1999. As pointed out in the US EIA's *International Energy Outlook 2000*, this price projection is higher than most, but is exceeded by the *IEO*'s own reference case by 2010. By 2015, following peak conventional production, the IEA's price assumption is higher still, over $30/bbl (1998$). Although spot prices did indeed reach this level in 2000, this is three times the low price of Brent crude at the end of 1998 and double the price at which a number of companies currently (mid-2000) test the economics of their future development projects.

There is also a further 'wedge' between the price of crude oil and the price of refined products or alternates: part of this is due to processing, transportation, refining and marketing costs, but in Europe and Japan a very large part is due to excise taxation by consumer governments. A barrel of crude selling at $20 provides a substantial profit for a typical producing state, but this price is a lower order of magnitude from that paid for the same volume of gasoline by the UK consumer, where a price of $200/bbl has been reached.[16] Note that UK, and European taxation generally, differs markedly from the US in both scale and trend (see Table 3.1).

The point that should be clear from this is that governments, through their taxation of upstream production and of retail products, exert an influence over supply, demand and price that has a fundamental impact on the ultimate level of reserves.

[16] The price of a litre of unleaded petrol on UK forecourts rose to 85.9p in June 2000, equivalent to over $200/bbl. Although this simple conversion ignores the fact that the crude barrel also comprises lower-value components, it does serve to illustrate the overwhelming share taken by tax.

Table 3.1: Gasoline taxes in Britain and the United States

	% of taxes in petrol prices, 1999	Change in real tax component, 1991–99 (%)
Britain	81	63
United States	28	–21

Note: Percentages relate to unleaded gasoline (95 RON).
Source: IEA, Energy Prices and Taxes, Q4 1999.

Time

Robert Mabro points out that perceptions of glut are more common than scarcity and have as much to do with short-term effects in oil market and general economic conditions as they have to do with the actual situation.[17] The first principle in thinking about this is to forget the short-term market, and focus instead on what the geologists and economists are saying. Figure 3.4 shows rising production (currently averaging 26 Gb p.a. and expected to rise further) set against discovery rates that have been falling for a long time. Or so it seems: some words of caution are needed. The discovery rates shown in the figure are five-year averages of those made in each time period, with any subsequent additions to these fields backdated to the date of discovery. These estimates are not of proved reserves (i.e. those that have a high degree of certainty attached to them); they are p50 estimates, which by definition are just as likely to be too high as too low. Nevertheless they are rooted in time: these are industry estimates made at the turn of the century, based on the application of known technology and assuming a modest price level. With changes to these assumptions – particularly relating to what future technology will permit – and allowing more time better to appraise these discoveries, even these p50 estimates can be expected to increase in the future.

Economists are not concerned about absolute volumes, but see time as the crucial element. Provided reserves are not exhausted during the limited timeframe in which they have economic value, they are, in effect, 'infinite'. Technical progress also marches with time, and there remain unconventional sources and substitution. While it would be

[17] See e.g. his review article of Petroconsultants' 1995 study *The World Oil Supply 1930–2050* in *The Journal of Energy Literature*, June 1996.

Figure 3.4: World oil discoveries and production: five-year averages

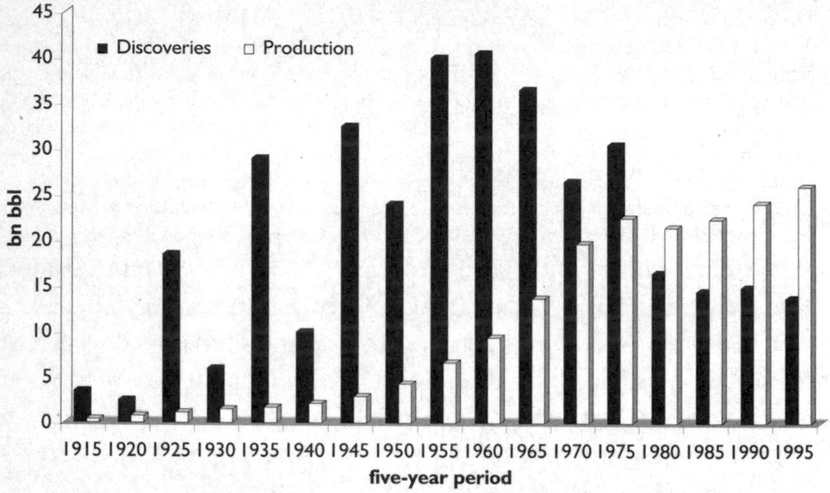

Sources: USGS, Petroconsultants/IHS Energy and BP Amolo *Statistical Review of World Energy 2000.*

wrong to have blind faith in technology, the economics and politics of investment are critically important, particularly investing in new capacity at the right time. The paradox at present is that the private companies are most active in those regions where oil is relatively scarce, but less able to participate where it is more plentiful. Hence the key question is: will Middle East OPEC invest in time?

Unconventional oil

Unconventional oil may be described as all oil that does not flow from reservoirs using traditional technology. Together, future reserves of Canadian bitumen and Venezuelan heavy oil could exceed 1,000 Gb, according to the US Department of Energy.[18] Shale oil resources worldwide (see Box 3.3) are estimated to be 15,000 Gb. The IEA's definition of unconventional oil also includes synthetic crudes, products derived from oil sands, and liquid supplies from coal, biomass and gas. Despite these massive volumes, the world's current unconventional production

[18] EIA, *International Energy Outlook 1999*, p. 23.

Box 3.3: Shale oil

The shale oil resource is vast, and is estimated in trillions of barrels. Shell estimates the better deposits to be 3800 Gb in place, with RFs as high as 50–80%. While widely distributed, much of it is concentrated in basins in the heart of the United States. Although small-scale oil shale industries have operated elsewhere since the nineteenth century, it was the peaking of US oil production in 1970, and the perception of future shortage that led to the formation of large oil company consortia to develop the mid-continent resource – all without success. Why? One reason is that the beds are buried deeply and only a few are thick enough to mine efficiently. Another is that oil shale is not oil: it does not flow, but needs to be drilled and blasted, and it isn't chemically oil; the organic material in oil shale is kerogen, 'raw' oil that has not been 'cooked' naturally to form petroleum. Once extracted it needs to be transported to a processing plant, crushed and heated. The missing hydrogen must be added, requiring large quantities of water before the standard refinery process can make the petroleum products that people want. There is also a waste disposal problem. This whole process means there is limited net energy gain.

Companies are, however, currently developing this technology. For example, the Stuart project has recently been built, a $250m shale oil pilot plant situated near Gladstone in Australia. At present it remains to be shown how development costs (which may be over $20/bbl) can be brought down to make other projects attractive at typical oil prices, and how this type of development can provide better returns than *in situ* pyrolysis. Even if this particular project is successful, it is unlikely to signal the go-ahead for US shale-oil development, since deposits there are generally of lesser quality.

Source: Walter Youngquist, 'Shale Oil: The Elusive Energy', *Hubbert Center Newsletter* no. 4 (1998).

is a mere 1.5m b/d. With the exception of South Africa's oil from coal (150,000 b/d) and the rest of the world (100,000 b/d), all production is in the Americas. Generally, the high cost of extraction limits in reserves but there is a supply curve, depending on the properties and location of each deposit. To allow for this, the IEA increases its price assumption from $17 to $25/bbl (1990$) as the world's supplies of unconventional oil are increased to replace declining conventional supply in the period 2010–20. It notes that 'advances in technology may well limit the extent of such a price rise' but that 'more stringent environmental controls in the future could increase it'.[19] Increased usage of unconventional crude is likely to result in slightly higher carbon emissions, although synthetic products will be cleaner.

[19] IEA, *World Energy Outlook 1998*, p. 113.

To date, most unconventional oil comes from steam injection and the mining of tar sands in Canada – where unconventional oil now already accounts for a significant fraction of national production – and from Orinoco heavy oil in Venezuela. In these countries production methods are competitive with conventional oil today.[20] Shell's Muskeg River project, for example, is projected to be economic even with prices at $10/bbl. In Northern Alberta alone, at least $15bn is currently earmarked for further development of tar sands, with the result that Canada's production of unconventional oil could rise from roughly 0.6m b/d to around 2m b/d by 2010.[21] Technological progress may well allow oil to be produced from other unconventional sources in the future without raising the price.

A switch to heavy and unconventional oil as the main source of new supply will mark a new era for the oil industry and for oil policy. Since the 1970s the private sector of the industry has been driven to find new oil in the areas opening up to it and then to increase the economic rate of recovery of the oil found. The low-cost state producers in export-dependent countries have been driven by the need to raise revenue, by either expanding or constraining supply. By 2010–20, or soon after, the picture will change: the private sector will be concerned with the development of known unconventional reserves. Here the question is not the size of the resources in total but the costs of developing particular deposits. The economics of unconventional oil will be conventional: competition from unconventional oil will be like competition from non-OPEC oil in the 1980s and 1990s. For the exporters of low-cost conventional oil, the situation will also change: as each exporter approaches the limits of its capacity to expand production, the question of holding back production to protect the price becomes academic. They will benefit from their lower costs, but will no longer need production quotas to restrict the competition among themselves.

[20] See e.g. Larry Fisher, Louise Gill and Ken Warrington, *Supply Costs for Canadian and International Crude Oil Sources*, CERI Study no. 88 (Calgary: Canadian Energy Research Institute, June 1999).

[21] *Petroleum Intelligence Weekly*, 26 June 2000, p. 3.

Policy matters

Political, environmental and strategic considerations will also influence the future of oil supply. Governments in most of the main exporting countries control the level of activity of state oil companies through their budgets. As mentioned above, all governments have a controlling influence on levels of industry activity, even where the energy sector has been privatized. This is exercised through legislation relating to, for example, the granting of licences for continued exploration and development, tax levels and environmental obligations.

Environmental and socio-political issues

Oil has a dual image. Its positive side is usually dominant, as users rely on oil's refined products to power transport and heat or cool homes as required. The negative side of crude oil is seen by most only when accidents occur – when it spills from ships, pollutes beaches or kills wildlife. Environmental pressures arising from the effects of urban pollution and concern over global warming have already resulted in the setting of international CO_2 emission targets in the 1997 Kyoto Protocol. Environmental arguments are used to justify the high levels of fuel taxation that already constrain oil consumption in most developed countries. To many, these pressures appear more likely to limit ultimate hydrocarbon usage than the size of the resource base itself. Indeed, Greenpeace has argued that stabilizing atmospheric CO_2 at 350 ppm (close to the current level) means that the world can only make use of one-fifth of current hydrocarbon (including coal) reserves.[22]

In addition to global climate concerns there are local environmental and habitat issues in certain areas, e.g. the Atlantic Margin in the UK; Alaska and offshore areas in the United States. A good example of the salience of such issues was the motion raised at the BP Amoco shareholders' meeting in April 2000, objecting to the development of the Northstar field in the Arctic.[23] Continental Antarctica is another area

[22] Bill Hare, *Fossil Fuels and Climate Protection: The Carbon Logic* (Amsterdam: Greenpeace International, 1997).

[23] This motion failed, but nevertheless attracted 13% of the vote.

from which exploration is excluded, although large reserves are not expected to lie beneath it. Ecuador has recently barred all industry activity in certain areas. All round the world governments are enacting legislation designed to protect the local environment by setting stringent standards under which the industry must operate.

Pure environmental issues tend to become tied up with local social or political issues. Opposition to oil companies in Ecuador, for example, is not simply a matter of fear of local pollution, but is wrapped up with the belief that the profits of extraction end up with elites in distant cities. Such a mix of environmental and socio-political concerns has also been evident in the history of Shell's operations in Nigeria. In other countries companies also become involved and implicated in more general human rights issues. This may occur at a local level (e.g. the hiring of security forces to guard particular operations) or may extend to any dealings with certain governments, such as the present regimes in Burma and Sudan.

Principal exporting countries

The key to future world oil supply lies in the hands of the state oil companies of a few exporting countries. The 10 largest private companies produce less than 20% of the world's oil. Although private sector companies invest over $100bn annually in exploration and development, this investment is largely targeted on areas outside the Middle East, away from this focus of low-cost oil-field development. A key structural question for the future is whether major exporting countries which now rely on their state enterprises to develop future production will seek to attract a larger share of this flow of capital and the associated technology and management which the private sector brings.

The OPEC countries of the Middle East hold over 60% of the world's proved reserves, yet supply less than 30% of total production from less than 1% of the world's producing wells. Finding and development costs are low – so what prevents accelerated production from this area?

The opening of *Iran* to foreign oil industry finance and technology is currently taking place. *Iraq* has a vast, low-cost, undeveloped reserves base, capable of producing 8m b/d by 2020. Current production of about 3m b/d is constrained by UN sanctions. In *Kuwait* a limited number of foreign consortia will be invited to bid for development projects. The only Gulf country that is not likely to invite companies into upstream oil imminently is *Saudi Arabia*. With around 2m b/d spare capacity, there is no need to do this at present.

Twenty years on, production from OPEC Middle East had not regained the levels achieved in the late 1970s, before the second price shock, the Iranian revolution and the Iran–Iraq War sent production sliding. The differences between these states that again led to war in 1990–91 are likely to cause friction again in the future. A substantial rise in Middle East production thus cannot be taken for granted. Indeed, it does not feature in some scenarios.[24]

What Middle East OPEC produces, as the world's residual supplier, also depends on the supply situation elsewhere and the level of prices. The trend of future conventional non-OPEC production is unclear:[25] there has been a levelling off of the rate of increase in non-OPEC supplies in recent decades, and it is not clear how long increases will continue before the production profile turns down. This depends to some extent on price levels. With higher prices, non-OPEC production can continue to grow for longer. It is only if prices are lower that the world becomes more dependent on the Middle East to balance supply and demand.

Norway's oil and gas industry suffered under the 1998 price crash. Investment in 2001 is likely to be only half 1998's $10bn. Current oil production is near its peak. In the long term, Norway's reserves make it more of a gas than an oil exporter. Recent cooperation with OPEC stemmed from a desire to defend prices to protect jobs in the industry. This motive weakens if prices rise above $20/bbl.

Russia and the rest of the FSU produced 12.6m b/d in 1988. Ten years on, FSU production had fallen to 7.4m b/d. The devaluation of

---

[24] See e.g. N. Nakicenovic, A. Grubler and A. Mcdonald, *Global Energy Perspectives* (Cambridge: Cambridge University Press, 1998); also the same authors' article in *Energy Policy*, vol. 27, no. 5 (1999), pp. 267–84.

[25] The US Department of Energy (*International Energy Outlook 2000*) has rising non-OPEC production and puts Middle East OPEC production, including NGLs, at less than 42m b/d by 2020. In contrast the IEA has this figure at almost 49 mb/d with conventional non-Middle East OPEC oil production *declining* in the same period by 1% annually.

1998 has improved the economics of domestic oil producers, but foreign capital has not flowed to most of the major new field and reservoir development projects. At present there are fundamental problems afflicting the Russian oil scene: the fiscal regime is complex and inflexible; too many taxes are based on gross revenue, not profits; administration is poor and political risk high. This situation should improve in the future. The recent discovery of a giant field in the North Caspian and the prospect of further development of Caspian reserves already provides hope that production in this region will grow. A gradual rate of growth is likely, determined not by resource constraint but by the political situation in the area.

The history of *Venezuelan* production shows that it is possible to reverse production decline, provided reserves are in place. Production peaked in 1970 at 3.75m b/d but subsequently declined, along with a loss of market share. Thirty years later, following the successful introduction of foreign capital, the upstream industry is set to exceed this rate and push towards a capacity target of 5.8m b/d by 2009.[26] The production constraint of 1999 may be seen as a temporary halt, but was also a principal cause of that year's remarkable price rise as the new government restored OPEC credentials.

Conclusions

Conventional oil resources are finite. Discoveries of new fields are not replacing what we currently extract, but the fraction of discovered resources that is being extracted – the reserves – is increasing. At some point in the life of a reservoir production will decline, and at some point in the life of a region the sum of these declines will outweigh the sum of the new discoveries and increases in recovery. Eventually the world may replicate the declines of the United States, its most mature oil producing area. Exactly when this will occur is uncertain; the IEA expects it to happen in the period 2010–20.

However, unconventional oil resources – heavy oil, tar sands and oil shale – are known to be very large, with quantities of oil in place many

[26] Reported in *Petroleum Economist*, July 1999, p. 34.

times those of conventional oil. Key questions relate to the environmental impacts and the economics of extraction and conversion to oil products or synthetic crude. The development of these reserves is likely to follow a supply curve starting with present production, and growing as costs reduce, environmental legislation permits, and the difficulty of extracting yet more from conventional sources increases. This transition to unconventional oil is already taking place in Canada, and can be expected to gather momentum, especially during periods of high prices, and become a feature of the period 2010–20.

About two-thirds of current world exports originate in countries where production is owned or closely controlled by the state, and where oil export revenues and rents are critical to the national balance of payments and budget. On mainstream present projections of demand and supply, most or all of these countries, principally in the Middle East, will reach peak production by 2010–20, or soon after. At that point, with no further need to limit competition among themselves, rivalry is likely to intensify. Additional competition from unconventional oil and other energy sources, from different countries, with private investment, is likely to limit price increases from 2020 onwards just as conventional non-OPEC oil did in the period 1980–2000.

Key uncertainties include:

- the timing of peak conventional production (though 2010–20 appears the key decade);
- whether politics will permit the increase in Middle East production that economics suggest should happen; and
- the shape of the industry in 10 and 20 years time. The rate of change is accelerating, but uncertainties abound. For the state companies, the national imperative to support growing populations while seeking to diversify the economy and develop industry will force changes to be made. The private sector will be challenged to meet shareholders' expectations while redefining the limits and linkages of their businesses to the financial, utility and vehicle manufacturing sectors and responding to increasing social and environmental demands.

It may be that oil production will peak not because of natural depletion of the resource base itself but as a result of environmental policy and/or falling demand, due in part to the replacement of oil by natural gas or the emergence of alternative fuels for vehicles. The balance of evidence suggests that resources will not be a key constraint on world oil demand in the period to 2020. As the EIA states, 'more important are the political, economic, and environmental circumstances that could shape developments in oil supply and demand.'[27]

[27] EIA, *International Energy Outlook 2000*, March 2000, p. 32.

Chapter 4

Transport in transition

Current trends in transportation are unsustainable. A struggle for new means of mobility is beginning. Changes are inevitable. Soon after 2020, if not before, growth in the supply of fuel for vehicles will depend on increasing competition among technologies for recovering liquid fuels from natural gas or from 'unconventional' oil. To manufacture clean petroleum fuels to protect urban air quality against increasing volumes of vehicle traffic increases in hydrogen inputs will be necessary; fuel and related engine developments will be required. Refining processes that increase the hydrogen content of fuels add to CO_2 emissions. The transport sector will be required to contribute to the reductions in greenhouse gas emissions required under the Kyoto Protocol. Even if the Kyoto target reductions in greenhouse gases (GHG) are not met, policies can be expected to push in that direction. This will involve going beyond the 20–30% improvements envisaged by current voluntary commitments such as those by the Association of European Automobile Manufacturers (ACEA) in Europe, since such improvements extend trends which are already embedded in 'business as usual' projections.

Moving beyond the voluntary agreements and similar arrangements will involve some combination of policies and changes in the marketplace. Taxes on transport fuels may rise – at least where they are now low – to incorporate the external costs of meeting the demand for mobility and influence the demand for future fuels and vehicles. Governments may dedicate public sector transport fleets (as is now done in the United States) to fuels requiring new distribution systems, e.g. compressed natural gas, or natural gas converted into diesel fuel and ethanol from biomass. Industries may develop new forms of engine less dependent on carbon-intensive fuels. Hybrid electric vehicles are likely to be widely available in the OECD countries by 2010. These do not require new fuels or distribution systems. Some vehicle manufacturers have announced plans for marketing vehicles powered by fuel cells. These require hydrogen to be supplied by new fuel conversion and distribution systems. Their effect on CO_2 emissions depends on the source of hydrogen.

Alongside fuel and vehicle technologies, new communications and signal control technologies will enable the installation and management of existing road and traffic systems (including road user fees) to reduce congestion and improve travel speeds relative to fuel use.

In the long term, investment in rail and alternative forms of mass transport may change the relationship between economic development and the demand for transport. The traditional transport and traffic system will face competition from alternatives connected with changes in the shape and organization of human settlement. These are inseparable from deeper changes in the size, scale, nature and character of cities as urbanization intensifies. As these changes develop in industrial countries, differences will grow up between the fuel and vehicle industries of industrial countries and those of developing countries where transport based on petroleum-fuelled vehicles will continue to grow very rapidly, serving and reflecting economic growth.

By 2020 there will be a variety of models of transport competing with transport dominated by petroleum-fuelled cars and trucks. The share of value added in transportation will shift from the primary energy source to the manufacturing process, from fuels to vehicles, and from cars and trucks to other forms of transport and communication. Value added will follow the new technology pouring into the new vehicles and systems.

Demand for transport fuel

About 18% of the world's energy is consumed in transportation, up from 11% in 1980.[1] Transport use tends to grow with GDP, while energy use

[1] EIA, *International Energy Outlook 2000*, p. 241.

grows more slowly. Most baseline forecasts by the IEA and the EIA[2] project that the transport sector's share of total energy requirements will increase by some 2–3% by 2020. Transport's 20% share of fossil fuel use will grow more slowly. Fuel for air transport is projected to grow at 3.7%, faster than for transport as a whole. Air transport now uses about 10% of all transport fuel. By 2020 this will grow to 15%, as it is expected to account for almost 40% of the total increase in transport fuel use.[3]

Table 4.1: Growth rates 1995–2020 (%)

	World	OECD
GDP	2.8	2.3
Fossil fuels	2.2	1.2
Mobility = transport fuels	2.5	1.5
Total energy	2.1	1.1

Source: EIA, *International Energy Outlook 2000*. IEA projections in *World Energy Outlook 1998* are similar (see Chapter 2).

Oil and the transport market

Oil provides about 95% of the fuel used in transport worldwide. Transport uses just under half of today's oil supply. Most of the balance is sold in markets where oil competes with electricity, fuelled by nuclear, gas, coal or renewables, or directly with other fossil fuels such as gas and coal. At the cost of additional refinery processing (and CO_2 emissions) the residual fuel oil which may enter the power market can be converted into components for blending into transport fuels. The prices of oil products for the transport markets are therefore indirectly constrained by competition from other fuels in other energy markets. What these projections suggest is that this constraint is likely to continue in the next 20 years. Beyond that time, if demand for transport were to outgrow oil supply, transport users could potentially outbid

[2] The IEA *World Energy Outlook 1998* 'business as usual' case has the percentage of primary energy attributable to mobility rising from 18% in 1995 to 20% in 2020 (excluding non-OECD combustible renewables and waste). EIA projections are similar.

[3] EIA, *International Energy Outlook 2000*, pp. 243–5.

other sectors: oil prices could lift to a new level set by the value of oil in transport. Many governments of oil-consuming countries currently collect by gasoline and diesel taxes a large part of the difference between the value of oil in transport and its value in markets where it competes with other fuels. If oil did not face that competition, crude prices would rise and government consumer taxes would be squeezed. This paradise for oil exporters, and painful purgatory for consumer taxation, is just below the horizon, but may remain there.

Under IEA and EIA baseline projections, by 2020 the transport market, and the refining processes necessary to supply it with clean fuels, could reach 55m–65m b/d under 'business as usual' projections. These numbers are compared with the 'conventional vision' of oil use according to the EIA (*International Energy Outlook 2000*; hereafter *IEO*-00) and IEA (*World Energy Outlook 1998*; hereafter *WEO*-98) in Table 4.2.

Table 4.2: Oil demand and demand for transport fuels (mb/d)

	1998	2020	Increase 1998–2020
Total oil demand	74	110–115	40–45
Oil demand for transport	35	55–65	20–25

Sources: EIA, *International Energy Outlook 2000*; IEA, *World Energy Outlook 1998*; EC (POLES).

About 70% of the difference in the range of demand forecasts for 2020 is accounted for by the EIA's high projection of transport demand for oil in North America. Either way, transport accounts for about half of the projected growth in oil demand and its share of total demand would just top 50% by 2020. Another way of looking at these quantities is that even if there were no increase from today's level of oil production, there would be enough oil supply in 2020 to cover the higher projected oil demand with about 13% 'over' for other uses specific to oil such as lubricants. Against the IEA (1998) conservative projection of conventional oil production (90m b/d[4]), transport demand would leave almost 30% in the market for other uses, competing with other fuels. There is controversy over all projections of supply beyond 2010 (see Chapter 3).

[4] Including processing gains: IEA, *World Energy Outlook 1998*, table 7.21, p. 120.

Higher conventional oil supplies may emerge, as the US EIA projects and many economists, sceptical about the reality of resource constraints, predict. The IEA's *WEO*-98, however, projects 'conventional' oil supply falling back to the 1998 level by 2020, with the growth in supply coming from liquids produced as a by-product of natural gas extraction, and from 'heavy' or 'non-conventional' oil, mainly in Canada and Venezuela. Beyond 2010, only about half this growth is covered by trends in the production of liquids associated with gas production and by identified other sources and projects.

Table 4.3: Oil supply, 1998–2020 (mb/d)

	1998	2010	2020
IEA oil supply	74	95	112
Conventional	64	79	72
Non-conventional	10	16	21
Unidentified	–	–	19

Since 1980 'gaps' in projections of oil supply have in practice always been filled by the development of additional conventional oil supply. As the boundary between 'conventional' and 'non-conventional' supply becomes blurred, the question will arise whether different new types of liquid supply can be matched to different markets. All 'unconventional' supplies will aim at the transport market directly, if processes exist which avoid production of by-products for lower-value markets – for example by converting methane from natural gas. Where this is not possible – for example, in processing 'heavy' (high-gravity) oil – the by-products will continue to compete with oil through the non-transport markets. In this changing pattern of competition, technology will play a critical role in product development and differentiation and in sustaining the growth in supply of fuel for transport beyond 2020.[5]

[5] Venezuelan 'heavy oil' can be converted by energy-intensive refining processes. There are competing processes for converting methane from natural gas in diesel and related components. The resulting product would be free of sulphur and thus have a premium value comparable to diesel from most types of crude oil. See 'Shell in Talks to Build Gas Plant', *Financial Times*, 30 June 2000.

Figure 4.1: Transport fuel versus GDP per capita

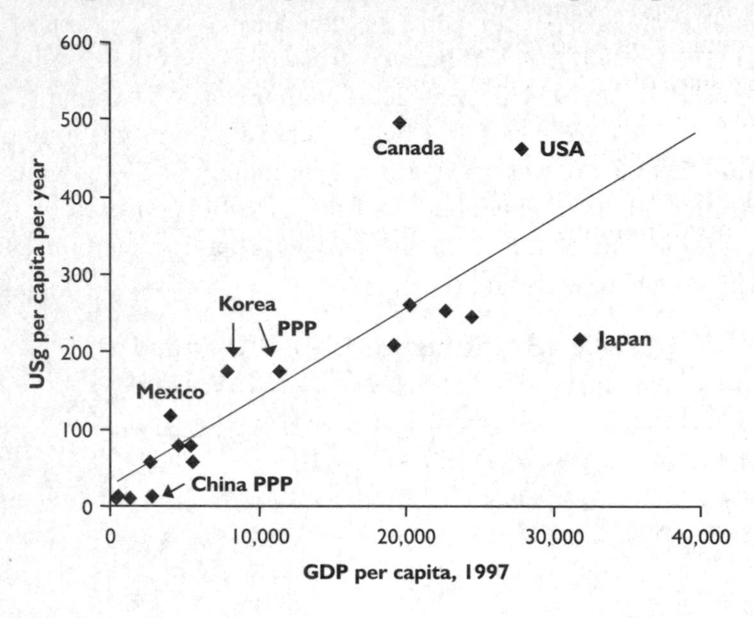

Transport demand

The demand for mobility, as expressed in the demand for transport fuels, is linked very strongly with the level of economic development as reflected in income per capita. This is illustrated in Figure 4.1, which compares the use of transport fuel per head with GDP per head for different countries. In the long term the price of transport fuels and the total cost of transport have an influence. This is shown by the outliers on the graph. Gasoline prices in Japan have since 1980 been consistently about double those paid in the United States or Canada,[6] mainly owing to higher taxes on gasoline.[7]

How the demand for mobility translates into demand for fuel depends in the short term on the legacy of vehicles and transport systems. Vehicles may reach the end of their normal economic life in 10–15

[6] IEA, *Energy Prices and Taxes*, fourth quarter 1999, part II, tables E8–E11, pp. 343–6.

[7] The dependence of transport fuel demand on income, with the long-run influence of price, is shown in a number of studies. For a discussion see IEA, *Transport, Energy and Climate Change* (Paris: OECD, 1997), pp. 42–6.

years, buildings in 20–50 years, highway and rail systems in 50–100 years. Scrapping capital in response to changes in the price of fuel or new regulations is costly and unlikely. In the medium term, new technology may offer less energy-intensive mobility. In the long term the demand for mobility may be affected by changes in the geography of living and working.[8] The demand for mobility, and transport systems and fuels, also respond to social needs for clean air and, more recently, to desires to minimize climate change attributed to fossil fuel use.

Clean air and GHG emissions

Air quality has improved in most urban areas in developed countries,[9] despite increasing vehicle use. This process is likely to continue through to 2010 as older vehicles are scrapped and fuel and vehicle standards continue to tighten. Current and foreseeable maximum threshold targets for air quality and noise in developed countries in the period 2010–20 are likely to be met by applications of known technology under existing regulations or identified trends in regulations.

Meeting air quality standards will, however, tend to increase CO_2 emissions. To make the cleaner fuels, refiners must use more hydrogen.[10] Supplying hydrogen either from within the oil-refining process (by partial oxidation of vacuum residue) or by converting natural gas (steam reforming) also generates CO_2: about 14.5 tonnes and 10 tonnes of CO_2 per tonne of hydrogen respectively. The net effect is that projected transport fuel trends for 2010–20, in developed countries, cannot be met by clean fuels and combustion engines without increasing CO_2 emissions per unit of transport fuel – perhaps by as much as 5%.

[8] L. J. Schipper and C. Marie-Lillieu, *Carbon Dioxide Emissions from Transport in IEA Countries: Recent Lessons and Long-term Challenges*, IEA Working Paper, February 1999, figure 9 and discussion.

[9] Ibid., figure 3.

[10] Several processes are involved. Converting residues into clean components generally requires additional hydrogen (for example, hydro-desulphurization); excluding certain components from the gasoline or diesel pool cuts out processes which potentially generate hydrogen (e.g. catalytic reforming).

The Kyoto Protocol and after

Under the 1997 Kyoto Protocol, governments of industrial countries are committed to reduce the emissions of all GHGs, including CO_2, by the period 2008–12. For 2010 this would involve reductions of the order of 20–25% below 'business-as-usual' projections of OECD fossil fuel demand. For 2020, if the Kyoto commitments for 2008–12 were simply extended, the reduction would be over 30%. Even if Kyoto commitments are not met (see Chapter 8) policies will tend in that direction.

To meet climate change policy objectives while allowing the growing demand for transport fuels to be met, governments would need to try to reduce growth in the demand for transport fuel itself, and/or to reduce emissions from other sectors sufficiently to accommodate the increased demand for fuel for transport.

Efficiency or equity

There are economic reasons why the transport sector should be exempt from penalties on CO_2 emissions. In 1997 (before both the oil price collapse of 1998 and the surge of 1999), $1 spent on gasoline generated about 700 g of carbon in Europe, while $1 spent on fuel oil for electricity generated 12 kg of carbon. In the United States, where fuel taxes are lower, more carbon was generated per dollar spent, whether for gasoline or fuel oil, but fuel oil expenditure was still over four times more carbon intensive than gasoline expenditure. Put another way, in Europe motorists were prepared to pay nearly $1.5 per litre for gasoline – nearly 20 times as much as power generators paid for fuel oil which contained the same amount of carbon.[11] This higher 'price' for carbon in transport must reflect some combination of the higher value transportation fuel has in use, and the lack of alternatives.

[11] In the United States, where different fuel taxes and prices apply, $1 spent on gasoline generated nearly 2 kg of carbon, but $1 spent on fuel oil for electricity generation generated over 7 kg of carbon. In 1998, with low oil prices, carbon per dollar increased for US gasoline and for fuel oil in both Europe and the United States. For gasoline in Europe, rising taxes offset the fall in the oil price.

Table 4.4: Carbon prices implied in gasoline prices

	Gasoline	Heavy fuel oil
Carbon (kg/litre)	0.66	0.8
OECD Europe		
Price per litre (1997$, PPP)	1.00	0.07
Carbon (kg/1998$)	0.7	12
USA		
Price per litre (1997$, PPP)	0.37	0.11
Carbon (kg/1998$)	1.68	7.27

An efficient policy on CO_2 would provide incentives to reduce fossil fuel use where it has least value and where lower-carbon and cheaper alternatives are available: that is the basis of the arguments for tradable emission permits. There is no practical way to ration carbon to individual vehicle owners through emission permits, but if there were, the transport sector would clearly outbid all other sectors. Policies to reduce emissions in the most efficient manner should recognize this. There are avoidable economic costs in subjecting the demand for oil in transport to the same CO_2 percentage reduction targets as fossil fuels in sectors where carbon emissions carry less economic value.

Easier treatment for the transport sector would also be consistent with Greenpeace's idea of managing a 'carbon budget' for the planet:[12] if carbon (for emission) is to be budgeted, tonnes should be allocated among users so that users receive something like equal value. However, equity suggests that each sector – or rather, those who have assets committed to it – should 'bear its share'. The question is, 'Share of what'? If transport were to escape the costs of the changes necessary to impose new GHG reductions, the users of transport would enjoy a 'windfall' relative to the owners of assets in sectors where emission restrictions or taxes were imposed and costly changes were necessary. A tax aimed at this 'windfall' would be temporary (since the 'windfall' exists only while old capital stock is in use) and would not be designed directly to achieve target emission reductions in transport.

[12] Bill Hare, *Fossil Fuels and Carbon Protection: The Carbon Logic* (Amsterdam: Greenpeace International, 1997).

In the real political world there is no sign that transport fuels will be exempt from climate change policies, or that they will be easy to target. Even in developed and rich countries, transport serves many purposes, some of them economic (bringing people to work, moving goods), some (like leisure journeys) 'pure consumption', while some (related to schools and hospitals) serve 'development' or 'quality of life' objectives. Increasing the cost of transport affects all types of journey equally, but the ability and willingness of different users to pay (or not) the increased costs is not equally distributed.

Conflicts in policy

Policies to reduce GHG emissions from transport will conflict to some extent with environmental policies intended to reduce urban air pollution and provide 'clean air'. GHG targets are unlikely to gain priority over clean air objectives: urban air quality is too immediate a problem, given the rate of growth of urban pollution and traffic. Clean air targets will not be relaxed, and in industrial countries will largely be achieved over the next decade. GHG objectives will add a new criterion for evaluating alternative transport. This blocks some air quality solutions (such as electric cars using coal-generated electricity).

New policy compromises

Countries are developing a barrage of climate-related policies for transport fuels, vehicles and systems on top of those already in place to improve air quality. Effectively to dam the growing flood of fuel demand requires policies for each stage. Price signals can influence demand by users. Investment incentives may induce the vehicle industry to develop new vehicles to provide more efficient (or less carbon-intensive) mechanisms for mobility. Investment (generally public) in infrastructure for alternative modes may increase the choice available. A planning framework friendly towards increasing urban population densities can help mass transit systems to become economic and relatively more acceptable.[13] Often there are gaps: the pieces do not fit together automatically.

Policies intended to reduce the demand for mobility

The econometric evidence is that the effect of fuel price, and therefore taxes, on the demand for mobility is only high in the very long run (as shown by cross-country comparisons). The UK 'escalator' of 6% per annum in real terms in the road fuels tax has already brought UK gasoline and diesel prices above those in most industrialized countries.[14] Moreover, in the EU and Japan fuel taxes are already very high, so that higher absolute increases are required than in North America to achieve the same reductions in demand.

Modelling studies suggest that user and access charges can reduce the demand for travel in towns, provided there are alternatives such as public transport.[15] If investment in new highways and parking space is withheld, there is rationing by congestion. This has an economic cost in wasted time and pollution cost in terms of avoidable emissions from idling vehicles. Model studies suggest that peak-hour charging and investment in traffic management can yield a 'double dividend': both more efficient use of the infrastructure and lower emissions.[16]

---

[13] The economics of bus, metro and related mass transit systems, and the attractions of cycling and walking alternatives, are very sensitive to high local population densities.

[14] The balance between taxes on diesel and gasoline illustrates a trade-off: diesel tends to be more energy- and CO_2-efficient, but to emit either more particulates or, in the case of 'city diesel' more NO_x than the gasoline alternative. In the United States, the Energy Policy Act of 1992 was designed to promote domestic alternative fuels in order to reduce energy imports: federal and state vehicle fleets are required to contain a rising proportion of alternative-fuelled vehicles. Alternative fuels include natural gas, coal-derived liquid fuels and electricity (which in the United States is generated from mainly domestic resources, over half of which are coal).

[15] See e.g. Ranjan Kumaar Bose and V. Srinivasachary, 'Policies to Reduce Energy Use and Environmental Emissions in the Transport Sector', *Energy Policy*, vol. 25, nos 14–15, pp. 1137–50.

[16] See articles by Roberto Roson and Peirson and Vickerman in Roberto Roson and Kenneth A. Small, eds, *Environment and Transport in Economic Modelling* (Dordrecht, Boston and London: Kluwer), pp. 39–60, 61–75.

Tax incentives work where alternatives are available; prices and taxes affect consumer choice of fuel and vehicles. This has been demonstrated in the switch to diesel passenger vehicles in Europe under the influence of tax advantages based on considerations of fuel efficiency and a desire to reduce oil imports. As more attention is given to the higher particulate emissions arising from the use of diesel, some of these tax advantages are being eliminated. This illustrates the priority for clean air over CO_2 and energy security objectives.

Policies to increase the supply of more efficient conventional vehicles

In OECD countries the supply of vehicles is the responsibility of the vehicle manufacturing industry, mainly in the private sector. Governments of the major manufacturing countries may influence future supply by direct negotiation with the industry, backed by regulation.

The US Corporate Automobile Fuel Efficiency (CAFE) standard, established under the 1980 Clean Air Act, brought about improvements of the order of 50% in the fuel efficiencies of broad classes of new road vehicles.[17] By 1990 US new cars were more energy efficient than comparable European cars, but Americans chose to buy bigger cars and drive more.[18] The question for the future is whether (or when) these standards should apply to off-road vehicles (now 30% of the household vehicle fleet) and whether a new generation of standards should be set.

The EU in 1998 secured a voluntary agreement with the ACEA to improve the average fuel efficiency for new cars sold by them in the

[17] See Schipper and Marie-Lillieu, *Carbon Dioxide Emissions from Transport in IEA Countries*, figure 19.

[18] Similar programmes in California to increase the proportion of sales of zero emission or ultra-low emission vehicles (ZEVs and ULEVs) have been less successful because of the difficulty of developing electric (battery-powered) vehicles with performance and range similar to gasoline-powered vehicles. California (and other states which had set such targets) have eased the quantitative target for the years up to 2003, and introduced the concept of EZEV (equivalent zero emission vehicles), whose emissions may equal those of the power plants which would have generated the electricity for a similar electric vehicle. Either the technical requirements could not be met or manufacturers could not develop models that California would accept.

EU in 2008 by 25% compared to 1995.[19] The European Commission has negotiated similar commitments from Japanese and Korean suppliers to the European market. There are parallel commitments by Japanese car manufacturers for sales in Japan. Such agreements allow manufacturers to choose how to meet targets by changing the designs of different components of their sales fleet. Although they would argue that they will choose a mix that 'consumers want', consumers' choice is in fact limited in the medium term by what is available from the long lead-time process of design and model development. Incentives specific to fuel efficiency and emissions might more fully exploit the technical possibilities available.[20]

Typically, automobile trade between the United States, Japan and Europe is in larger, higher-performance cars than the average of sales on the domestic market. Trade disputes therefore arise. The WTO has ruled, in the case of imports to the United States, that tariffs matching excise taxes on larger cars are non-discriminatory and consistent with the GATT, but requiring importers to match the Corporate Automobile Fuel Efficiency (US) (CAFE) standard of average efficiency for domestic sales is regarded as discriminatory.

Automobile fuel efficiency targets of this kind can probably be met until 2010–20 by the changes which the industry can identify[21] to existing types of vehicles through combinations of:

- improved combustion technology for gasoline and diesel engines (efficiency gains of around 30% are achievable by direct injection with stratified charges and variable valve timing, and many major manufacturers will be marketing direct injection propelled vehicles by 2000 – some are already on the market);[22]
- electronic clutch controls;

[19] European Commission press release, DN IP/98/734, 29 July 1998.

[20] For a development of this argument for Europe, see Per Kågeson, *The Drive for Less Fuel* (European Federation for Transport and the Environment, 2000).

[21] André Douard, '2020 Engines', working paper prepared for Transport Workshop, Royal Institute of International Affairs, London, April 1999.

[22] Diesel: Alfa Romeo, Mercedes, PSA, Renault, Isuzu, GM, Ford diesel vehicles, Mitsubishi, Toyota, Nissan, Renault, PSA, VW, and DB-Chrysler for gasoline vehicles. Source: Douard, '2020 Engines'.

- reducing friction by design and new lubricants;
- reducing weight through new materials, while maintaining safety through design changes;
- continuing improvements in aerodynamics; and
- matching these changes in the vehicle propulsion and transmission systems by developments in fuels and lubricants; by 2020 the new gains from these changes will be diminishing, and air quality problems will increase again because of the projected growth in the use of vehicles and the increasing degree of urbanization.[23]

Technological development

Development of new vehicle types

Governments, environmental movements and advocates of 'sustainability' all expect the vehicle manufacturing industry to be delivering new types of vehicles by 2010–20. Without any expectation of very strong government assistance from taxes or protected markets, the industry is developing models which will be close to existing vehicles in performance, range and price, unlike the 'zero emission vehicles' (ZEVs) – electric cars which were once mandated for California, but which did not materialize.

As well as greater fuel efficiency, these new generations of vehicles need to have lower GHG emissions, to satisfy requirements for near-zero emission of air pollutants such as volatile organic compounds (VOCs), nitrogen oxides (NO_x), sulphur oxides (SO_x) and particulates. Manufacturers will also be required to achieve lower noise generation per vehicle to protect urban environments as the number of vehicles increases, and provide for cheaper and environmentally less damaging recycling and disposal.[24]

---

[23] Peter Wiederkehr, 'Environmentally Sustainable Transport (EST), International Perspectives: OECD's EST Project', working paper for RIIA 2020 Project workshop, 2–21 April 1999, figure 4, p. 4.

[24] See criteria for OECD EST project: Peter Wiederkehr, 'Environmentally Sustainable Transport (EST), International Perspectives', presentation to OECD workshop on environmentally sustainable transport, Berlin, 21 September 1999 (Paris: OECD).

Some governments promote industry cooperation to achieve these goals. In 1993 the United States established a Partnership for a New Generation of Vehicles (PNGV) as a 10-year programme involving the administration, GM, Ford and Chrysler (now Daimler Chrysler). The objective is to develop a range of options for medium-sized family cars with 'three times' the fuel efficiency of conventional cars. In fiscal 2000 PNGV received $263m of federal funds. It is focused on four key technologies: hybrid electric vehicle drive, direct injection engines, fuel cells and lightweight materials.

Battery electric cars

These vehicles have evolved mainly to meet requirements in California and some other regions for zero emissions of local pollutants. Their range is limited to around 50 miles for lead acid batteries and 100 miles for nickel-based batteries. Depending on the fuel used to generate electricity in power stations, CO_2 emissions may be higher (in the case of coal-generated electricity) or lower (in the case of gas, nuclear or renewable electricity) than for conventional cars. Sales are aimed at the protected markets of state and local government fleets and local commercial hauliers in the United States and Japan.[25]

Hybrid vehicles

These vehicles work at high efficiency with low emissions because the combustion engine operates at a constant optimal speed,[26] generating power which is stored for use as required in the electric propulsion system. Hybrid vehicles have many advantages: their range is comparable to that of conventional vehicles; 20–30% efficiency gains are possible

---

[25] Examples of battery electric vehicles at or near commercial availability for protected markets are the Toyota RAV4 sports, the Ford Ranger, Dodge Caravan GM's EV1 and Chevrolet S-10, Nissan's Altra and Prairie. *National Alternative Fuels Hotline* <http:/www/afdc/doe.gov/hotline.html>, October 1988; *Toyota News*, 3 March 1999 (website), <http:/www/toyota.com/times>.

[26] If the engine is lean-burn it can also stop and start without deteriorating combustion. Additional power may be gained from regenerative braking.

(compared with direct injection gasoline or diesel engines); no energy is wasted idling; emissions are reduced because combustion is optimized; power from the combustion engine can be switched automatically between battery charging and transmission depending on the load of the combustion engine (especially lean-burn engines); and the combustion engine can be switched off when the vehicle is used in ZEV or high-pollution areas.

Hybrid vehicles using gasoline or diesel are beginning to reach major markets. One, the Toyota Prius, presented as a concept vehicle in 1995, was marketed in Japan in 1998, the United States in 1999 and in 2000 in Europe, where it faces competition from a Honda hybrid. By 2010 a variety of hybrids is likely to be on offer,[27] using gasoline, diesel or compressed natural gas (CNG) according to the availability and price of the fuel. Because their driving characteristics will be similar to those of conventional vehicles, how far they penetrate the market depends on price, which in turn depends on manufacturers achieving lower costs or governments introducing favourable vehicle tax treatment (or both). By 2020 hybrids could be a significant fraction of the fleet on the road.

Fuel cell vehicles

Fuel cells convert hydrogen to electricity which can be used to power stationary engines or vehicles. Recent progress in fuel cell technology, with the development of the polymer electrolyte membrane (PEM), has improved the efficiency of prototype fuel cells, and other options may emerge. Further cost reductions can be expected as a result of technical development and the experience of manufacturing fuel cells for stationary use, probably fuelled by hydrogen from natural gas. The question of hydrogen sourcing for vehicles is more open; in contrast to the hybrid vehicles, vehicles with fuel cells will require new distribution systems to be established unless they are fuelled by gasoline or diesel.

Electric powering of the vehicle probably offers efficiency gains of 20–30% (like the hybrids). If the fuel cell were powered by the

[27] GM has announced plans to market a hybrid in 2001.

on-board reforming of gasoline or diesel (a Shell process), it could refuel within the existing distribution network, but there would be local emissions to the air. Methanol (for on-board reforming to hydrogen) would require changes in the distribution system and might present health risks. (Methanol is being withdrawn as a 'clean fuel' in California because of risks of groundwater contamination from spills and leaks.) If the fuel cell were powered by hydrogen from methanol (the Daimler Chrysler NECAR concept), a fuel cell powered vehicle would qualify as a zero-emission vehicle for local pollution controls. Fuel cell vehicles (apart from the question of cost) would require the introduction of a new distribution system to supply the fuel and a source of hydrogen which (in the absence of new nuclear power) would involve fossil fuel burning. The use of ethanol from renewable sources (which could also be used on a combustion engine) would avoid the CO_2 effects of making hydrogen from fossil fuels, but ethanol could also be used directly in a combustion engine. There must, however, be doubts about the scope for developing adequate supplies of ethanol for a significant fraction of the demand for transport fuel. The net GHG effect of the adoption of fuel cell vehicles is currently under debate: much depends on the source of hydrogen and on further developments in the efficiency of the fuel cell.

Policies to promote modal change

Projections of 'business as usual' demand for transport fuels assume the continuance of existing trends towards increased urbanization. Since 1980, the urban share of the world population has increased by 5%; in China and East Asia by 10–11%. More than half the population of developing countries is expected to live in cities by 2015, and of 26 urban zones expected to contain more than 10 million people by then, only 5 are in OECD countries.

Urbanization is a complex phenomenon. The availability of transport between cities and between cities and food, water and mineral resources is also complex. The increasing weight of urbanization will weaken the link with the growth of transport fuel demand because urban air quality (and noise and safety levels) will deteriorate at health and safety costs which are avoidable by improvements in fuels, vehicles and alternative transport systems. The priority (described above) accorded to these improvements will not relax. Competition for space (and resulting congestion) will limit the rise in the number of vehicles used. With increasing densities of population, the economics and acceptability of public transport alternatives and mass transit will improve.

Other developments will be necessary to preserve and develop an acceptable quality of life in increasingly concentrated human settlements: changes in methods of restraining crime, in social conditions, in social support and in education programmes will shape the way cities work in the future. The general effects of cheap and accessible microcomputing and communications will allow much more efficient use of the space available. These 'clever' systems can be applied to existing infrastructure and building patterns. They will be echoed in the use of better information and communications to manage social infrastructure such as police, health and education services. Systems will be more generally available that were not at the disposal of policy-makers (even if there had been the will to use them) during the early 1980s when attempts were made to discourage the use of vehicles in towns for energy policy reasons.

By 2010 many large cities in developed countries are likely to be experimenting with some combination of these measures; by 2020 the more successful ones will be proliferating. These may include:

- charging for infrastructure within as well as between urban areas;
- charging for access to congested areas and at peak times (and increasing and differentiating parking charges);
- zoning of traffic with differential restrictions; and
- the use of car sharing or local pools of low-energy vehicles.

The effect of such systems is complex: to the extent that the efficiency of urban transport is improved, transport demand may be increased,[28] though with lower emissions of pollution and CO_2 than would other-

[28] Increased efficiency works like a reduction in price. See D. L. Greene, J. Kahn and R. Gibson, 'Fuel Economy Rebound Effect for US Household Vehicles', *Energy Journal*, vol. 20, no.3 (1999).

wise be the case. Depending on the detail, there could be 'double dividends' from better urban conditions and the economic benefits of transport usage, or 'double penalties' from commitment to poorly designed systems.[29]

Increased application of constraints and pricing to individual vehicles in urban areas over the long term will change the economics of bus, light rail and metro urban transport systems. The lower degree of individual choice which such systems provide compared to today's use of individual vehicles may not in fact be much lower than the congested, restricted, taxed and charged use of individual vehicles for short urban journeys. Depending on the fuel used – so long as it is not coal-fuelled electric power – urban mass transport systems have much lower CO_2 emissions per passenger or kg/km than individual vehicles would have, even under well-managed traffic conditions. By 2020 such systems may absorb much of the growth in transport demand in urban areas in high-income countries.

At present, it is difficult to say how the growth of the Internet and e-commerce will affect these trends. Travel patterns and goods delivery systems in urban areas may change. In rich countries 'discretionary' journeys may expand while 'necessary' journeys grow less fast or diminish – presumably with an effect on the price elasticity of transport. There will be a premium on transport systems that minimize personal travel time and have predictable times for the movement of goods and people.

Developing countries: the case of Korea

Developing countries account for about 60% (13–15m b/d) of the growth in transport fuel demand in 'business as usual' projections, a 20-year growth rate of 4%, compared to 5% in the past and 2% in industrial countries. As Figure 4.1 showed, rising incomes drive higher levels of transport fuel consumption in developing countries as well as in industrial countries.

The case of Korea is particularly interesting, with a rate of growth of car population and transport fuel consumption that has been probably

[29] R. Roson and K. Small, eds, *Environment and Transport in Economic Modelling* (Dordrecht: Kluwer, 1998).

the highest in the world. Between 1987 and 1997 the number of passenger cars in Korea increased by 800%. Transport fuel consumption trebled. GDP per capita increased by 83% (in real terms). The length of roads increased by 60% and the length of paved road doubled. At about 120 cars per 1,000 people, Korea's car intensity is approaching that of Japan.

The growth of transport fuel consumption in Korea would have been even greater without investment in a subway system (taking a million more passengers in 1997 than 1987). There was also an increase of 30,000–50,000 air passengers (500% over the period), the number of buses in service more than doubled, and the number of rail passengers increased by 60%. The demand for mobility was met by an increase in *all* modes of transport.

The growth of transportation both enables economic development and results from it. Both involve rapid urbanization. As noted above, in developing countries as a whole half the population will be urbanized by 2015 (compared with a third in 1990). Four billion people will live in cities of over one million people.[30] The health, economic and social costs of failing to achieve tolerable levels of air quality and mobility will be very high. Clean air objectives are thus appearing in major cities in developing countries. As in the developed countries, air quality objectives cannot be met simply through slowing the rate of growth of fuel demand by increasing energy efficiency. Where energy efficiency can be improved without major investment – for example, by better traffic management to reduce congestion and idling time – there can be gains both in reducing pollution and in increasing economic efficiency.[31]

Improvement in fuel quality (reduction of lead, sulphur, NO_x) in Mexico City, Seoul, Sao Paolo and Bangkok is urgent for obvious reasons of public health and welfare. Progress towards acceptable air quality levels will require further tightening of standards along the lines of those adopted in the United States, EU and Japan.[32] As in industrial

[30] World Bank, *World Development Indicators 1997*.

[31] For a study suggesting possibilities for Delhi, see R. K. Bose and V. Srinivasachari, 'Policies to Reduce Energy Use and Environmental Emissions in the Transport Sector', *Energy Policy*, vol. 25, nos 14–15 (1997), pp. 1137–50.

[32] EIII and EIV standards for automobiles and vehicle fuels are examples.

countries, this will increase refineries' demand for hydrogen, and increase the CO_2 emissions per unit of fuel in final use.

Each developing country is different from all others, but most face the transportation problems of developed countries in magnified form: urbanization, economic development, rising demand for transport and the need to reduce pollution in cities. Countries such as Korea may mitigate these by substantial investment in buses, highways and transit systems. The Korean vehicle manufacturing industry, like the rest of the manufacturing sector, is export-oriented and must produce vehicles with fuel efficiency and pollution control standards to suit its export markets. The protected, often state-owned industries of China, India, Indonesia and other developing countries are not in that position. There is a trade issue here. Could freer trade and investment give consumers in developing countries cleaner vehicles and fuels?

Air transport

Demand for jet fuel for air transport is projected to more than double by 2020,[33] bringing it to 9.5m b/d – over 15% of all transport fuel demand. Half of the increase of 5.5m b/d since 1996 will be in industrialized countries. Rates of growth in developing countries will be faster, with Asia consuming as much energy for air travel in 2020 as North America does now. Air traffic is expected to grow faster, the difference being accounted for by continuing incremental improvements in efficiency and air traffic management. Efficiency improvements to about 2015 can be identified in current trends and designs. Beyond that the relationship between traffic growth and fuel demand is more problematic.[34]

The impact of aircraft emissions on the climate is complex. Emissions from subsonic aircraft affect both the stratosphere and the troposphere. Gases (CO_2, ozone and NO_x) and aerosols have complex interactions. Some do not mix rapidly with other elements in the atmosphere and

---

[33] EIA, *International Energy Outlook 2000*.

[34] Joyce E. Penner, David H. Lister, David J. Griggs, David J. Dokken and Mack McFarland, *Aviation and the Global Atmosphere* (Cambridge: Cambridge University Press/WMO/UNEP, 1999).

(unlike CO_2) may have direct effects on regional climate. There are scientific and modelling uncertainties as well as uncertainties about the trends in aircraft design, engines and fuels, and traffic management and demand. An indication of the possible net effect is that according to the IPCC reference scenario IS92a, the share of global radiative forcing due to aircraft will grow from 3.5% in 1992 to 5% in 2050.[35] The balance of air travel, other forms of travel and other forms of communication is difficult to predict beyond 2020, but unless or until trends change, air travel will inevitably be a target for climate change policy.

Preparing for change

Industrial change

The vehicle and aircraft manufacturing industries are changing, with cross-border mergers, acquisitions and alliances reflecting the globalization of markets; the lowering of trade barriers; the globalization of capital and investor pressure for economic performance; and some reduction of government support for, and protection of, 'national champions'.

The changes in the outlook for the transportation sector as a whole compound the challenges facing firms in these sectors; vehicle manufacturing companies are likely to seek to share in the volume growth in demand in developing countries for vehicles of today's industrial country standards. They will also compete for a share, and if possible leadership, in adding value through the development of new generations of vehicles to meet the industrial countries' pollution and climate change objectives. There may be experiments to move beyond the manufacturing boundaries to provide more transportation-related services – including credit services – to consumers.[36]

The oil industry faces the same general challenges of globalization. In addition, most of the main private sector oil companies face a declining

[35] IPCC, *Aviation and the Global Atmosphere* (Cambridge: Cambridge University Press), section 4.8, pp. 8–9.
[36] See Tim Burt, Nikki Tait and John Griffiths, 'Ford's Full Service', *Financial Times*, 9 August 1999, p. 15.

upstream resource base, while the main public sector companies in developing countries are inhibited by state ownership and national boundaries, from taking part in the global restructuring of the industry.

In both transportation and oil industries, the changes under way open the door for new entrants to bring in new technology, resources or markets. Some of the alliances clearly focus on this: Shell has an agreement with DBB Fuel Cell Engines (whose shareholders are Daimler Benz, now Daimler Chrysler, Ford and Ballard Power Systems Inc.); Exxon has an agreement with Toyota (also allied since 1999 with General Motors for the development of battery electric, hybrid and fuel cell vehicles).

Policy issues

Transport is a target area for policies and preferences from many sources: consumers, firms, governments and organizations seeking to promote their environmental and social values through persuasion and direct action. Targets in transport move because the dynamics of the sector itself feed back constraints and new possibilities.

In the short term, economic activity and rising incomes drive the demand for mobility. Taxation will not change it quickly – certainly not quickly enough to halt or reverse the growth of fuel demand.

In the medium term, the types of vehicle available can change under the influence of competition among manufacturers, of taxation, and of regulations imposing standards on new sales or reserving 'niche' markets for alternative vehicles and fuels. Traffic management and charging schemes can probably reduce emissions and increase transport efficiency more quickly than alternative mass transit systems can be developed.

In the longer term, the transport sector can be further decarbonized by investment in different infrastructure, necessarily led by the public sector or protected in some way by regulation.

Integrating transport, energy and environmental policy is about addressing these three time horizons in a coherent way.

Chapter 5

Gas for oil markets

Introduction

This chapter looks at scenarios of global, regional and sectoral gas demand and prices.[1] It then reviews developments in gas markets in different regions of the world, highlighting four countries where the gas share of energy consumption grew particularly rapidly during the 1990s. A number of critical issues are identified for the further successful development of gas markets, and some conclusions are drawn regarding future challenges and uncertainties about how prices may develop along the value chain in different regions of the world.

The value chain

The value chain in the natural gas industry is different from that of oil. Because of the low density of natural gas in comparison to oil, a natural gas pipeline or container of a given volume will carry only 20% of the energy equivalent of the same volume of oil. This physical difference has a major impact on the costs in the natural gas value chain, in particular because most gas flows through pipelines from the point of production to the point of use.[2] Gas transportation costs – whether by

[1] This chapter includes material previously published in Koji Morita, *Gas for Oil Markets*, RIIA Briefing Paper, new series, no. 12, February 2000 (London: Royal Institute of International Affairs).

[2] The exception is the seaborne transportation of LNG by marine tanker, which can also be achieved on land by truck. The costs of pipelines are replaced by the cost of liquefaction, refrigeration of the liquid during transportation, and regasification at the market. Unlike marine transportation of oil, LNG transportation comprises less than 10% of global gas trade and allows far less flexibility of destination than crude oil and products trade. See James T. Jensen, 'The Development of Global Natural Gas Transportation Systems', paper presented at the Centre for Global Energy Studies, London, 22 June 1998.

pipeline or tanker – are much higher than those for oil, and can be multiples of the production cost.[3] For example, transportation comprises more than 90% of the total cost of delivering Siberian gas to west European borders through 5,000–6,000 km of large diameter pipeline. For gas fields in the North Sea delivering gas to the UK and continental Europe, the cost proportions would be more balanced, but for some Norwegian fields requiring long undersea pipelines at significant water depth, transportation would still account for the majority of the cost. Returns on the capital invested in transportation are a large component of the value added in the gas chain. The price of gas in consumer markets – especially in liberalized markets – is related (either by long-term contract formulae or by short-term competition) to the prices of alternatives: oil products such as kerosene and heating oil, or coal through electricity for heating and cooking. The high cost of gas transportation (compared to oil), varying by source, means that wellhead gas prices tend to be lower than their oil equivalents. The margin available for exploration and production investment is lower, and the profit and rent available for taxation are therefore also lower than is the case for oil. Since exploration costs are similar, upstream oil companies and governments tend to give priority to oil.

Natural gas is also a potential petrochemical feedstock – in the United States the main feedstock for ethylene manufacture. Where gas is located near petrochemical complexes (as in the United States and Scotland) it can enjoy higher netbacks. Where there are no economic alternative markets for the gas, petrochemical export industries may provide an outlet (as in Saudi Arabia, Trinidad and Venezuela).

Because the capital costs of building gas transportation (transmission and distribution) infrastructure are so large, in comparison to oil, gas prices differ substantially at different stages of the value chain in different places. Analysis of these costs and prices – and how they are calculated in different projects at different stages of the value chain specific to that project's outlets – is greatly hampered by commercial

[3] This generalization regarding costs is a little dangerous given the presence of certain impurities in the gas, and also the fact that significant gas is produced in association with oil where the treatment of cost is at the discretion of the producing company.

confidentiality. Lack of accurate historical knowledge in particular about costs of transmission and distribution means that trends are particularly difficult to forecast. The importance of this problem will become apparent later in the chapter.

The industrial structure

Since the 1970s natural gas has steadily grown in importance in OECD energy markets. This growth has been accompanied by great changes in the structure of the industry in most OECD countries. This process began in the United States with the Natural Gas Policy Act of 1978 (NGPA) and in the UK with the privatization of the gas and electricity industries in 1986 and 1990. In Europe the EU Electricity Directive of 1997 and the Gas Directive of 1998, though not changing the ownership of any part of the industries, are set to release the same forces of competition among suppliers for customers and among consumers for the choice and management of their supplies. The same forces are beginning to work in Japan as a result of deregulation there, and similar trends at work in other developed and developing countries will continue the process of change.

These changes, part of a general change in the relationships between government and economy, have some parallel in the changes imposed on the oil industry by the entry into the market of oil-exporting governments as owners and controllers of capacity they had nationalized or expropriated from the private sector. The similarity is that the changes in the oil industry structure, like those now moving across the world in gas and electricity, separated or 'de-integrated' parts of the industry. Trading between segments of the value chain released forces of competition which had previously been moderated by the dominance of integrated, global, multinational oil companies over all segments of international trade in oil.

There are, however, some differences that matter. One is the far greater importance of transportation in the value chain for gas than for oil, noted above. The second is the continued existence of an element of monopoly in some parts of the transportation chain. There is an almost inevitable monopoly in the physical system for the final stage of distribution of gas to consumers; but, as the British example has shown, with open access to this system there can be competition among suppliers even to household consumers. There is a temporary monopoly for a new long-distance transmission system until sufficient growth in demand occurs to justify the building of a competing system either over the same supply route or from a different supplier.

The third difference is in the role of governments. For oil, the governments of the main oil-exporting and export-dependent countries are, and will probably remain, responsible for key decisions about investment in new production capacity and about medium-term production plans. For gas, the situation seems different. Governments in Mexico and Saudi Arabia, which offer no opening whatsoever for private sector participation in investment in the oil resources of their countries, find it possible to create or invite openings into their (far less important) gas industries. In Russia, Gazprom, organized as a private sector enterprise, nevertheless has a 35% government shareholding and a special relationship with the government. There is a wide range of relationships between government and private sector producers in other gas-producing and exporting countries, including some, like Algeria, where the international private sector is a major component in the expansion of production and export. In some cases it is more difficult for foreign investors and state companies to agree to construct a production-sharing contract for gas than for oil.[4] While international market values are relatively easy to establish for oil, they are not so obvious for gas. Moreover, for many gas projects the first market to be satisfied is the domestic market, either for gas or for power. This can be difficult where prices for gas and power are controlled by governments at low levels. In such cases the markets are often supplied by an 'old style' (government-owned) monopoly. Monopoly transmission and distribution companies which could 'average in' the cost of new gas may still strain the tolerance of the price controllers. The experiences of Enron in the Daebol project in India, Shell in the Camisea project in Peru and the Malampaya project in the Philippines,

[4] For a description of the economic principles of production sharing agreements, see Kirtsten Bindeman, *Production Sharing Agreements: An Economic Analysis* (Oxford: Oxford Institute for Energy Studies, 1999).

and BP in Colombia reflect the difficulties encountered when a single project is required to bridge the gap between the 'old', local, price-controlled, monopoly power industry and the 'new', value-driven foreign investors in generation based on imported gas.[5]

For LNG, governments of exporting countries have in the past been able to take an 'oil-type' role upstream while leaving the transportation and marketing development to the private sector, which in turn has integrated and managed the risks of the value chain through long-term, take-or-pay contracts. As the markets at the consuming end of the LNG trade are liberalized – as in Japan and Korea – the basis for continuing such contracts will become more precarious.[6] As the LNG trade expands and involves more suppliers and more importers, a cargo or spot trade is beginning to develop, corresponding to the spot trade which has emerged in Europe and the United States.

In North America, gas has already become part of truly 'new economy' commodity markets for electricity, gas and pipeline capacity as well as for the commodities themselves. In OECD countries with established (and now liberalized) gas markets, gas may be ahead of oil in the development of sophisticated and interrelated trading mechanisms for transmission capacity, fuel and electricity. E-business can only accelerate these trends.

Resources and reserves

Some of the basic data on reserves, production and consumption by region are shown in Table 5.1. It is clear, even more than for oil, that there is a global surplus of gas reserves ready to develop and that there are great differences between the potential of different countries to supply. 'Proved reserves' of gas are over 60 times current consumption (compared to 40 for oil). The overhang of non-producing reserves has increased. Since 1970 reserves have increased by a factor of 4, while

[5] See EIA, 'Foreign Investment in the Electricity Sectors of Asia and South America', *International Energy Outlook 2000*, pp. 120–1.

[6] See Ulrich Bartsch, *Financial Risks and Rewards in LNG Projects (Quatar, Oman and Yemen)* (Oxford: Oxford Institute for Energy Studies, 1998).

consumption has increased by a factor of 2.2. As with oil, North American reserves have gone the opposite way, falling by nearly 30% over the period while consumption has increased by only about 10%. The importance of North American gas reserves has declined accordingly. North American gas reserves were nearly a quarter of the world total in 1970, but less than 5% by the late 1990s. In 1970 Middle East and Soviet reserves accounted for less than half of the total; by 1999 this proportion had risen to more than two-thirds. About 45% of the world's proved gas reserves are located in countries where current production rates are less than 1% of the reserves, and a further one-third of the world's reserves are in various parts of Russia, where current production rates are 1.3% of reserves.[7]

Table 5.1: Proven reserves of natural gas (1 January of each year) and production, 1999

	Reserves bn cu metres (% of total)				Production as % of reserves
	1970	1980	1990	1999	1999
North America	9,428 (23.9)	8,015 (10.4)	7,464 (5.7)	6,550 (4.2)	11
South America	1,874 (4.8)	4,353 (5.7)	6,921 (5.4)	8,296 (5.3)	1
Europe	4,053 (10.3)	4,497 (5.9)	6,044 (4.8)	7,871 (4.9)	4
FSU	12,086 (30.6)	31,000 (40.2)	52,000 (39.5)	56,677 (35.9)	1
Africa	3,834 (9.7)	5,683 (7.4)	8,490 (6.7)	10,444 (6.6)	1
Middle East	6,618 (16.8)	18,527 (24.1)	37,834 (30.4)	53,054 (33.7)	<1
Asia–Pacific	1,550 (3.9)	4,796 (6.2)	10,565 (8.9)	14,811 (9.4)	2
Total	39,443 (100)	76,871 (100)	129,318 (100)	157,703 (100)	1.5

Sources: Reserves: Cedigaz, *Natural Gas in the World, 1999 Survey*, table 2, p. 19; production: BP Amoco *Statistical Review 2000*.

This distribution sets the pattern for competition among gas exporters. The main medium for competition to expand gas markets will be the development of transportation for international trade; the main driver

[7] Figures from BP Amoco *Statistical Review 2000*, which reproduces reserves published in the *Oil and Gas Journal*.

will be the prizes to be won by connecting countries with large, surplus gas reserves to countries with potential for huge increases in gas consumption at costs which enable those transfers to take place. Large-scale projects will lower unit costs of transportation, but large grassroots projects require multi-billion-dollar finance which, in the past, has been achieved through pricing and throughput contracts made possible by an absence of competition in downstream markets. The prospect of rapid demand growth is encouraging the promotion of large-scale new transportation projects in many parts of the world. It is not necessarily very costly to propose these projects and keep them 'alive' in the queue of possible developments for particular markets; however, uncertainties about the timing of demand growth and future pricing conditions tend to deter the expenditure of serious amounts of money and the making of irrevocable commitments.

The combination of large reservoirs and markets with large and rapid growth potential focuses attention on certain links – from the Middle East to India, from East Siberia to China, and from Central Asian/Caspian countries to Europe. For western Siberian and Central Asian gas, all markets are distant, but reserves are large. The choice between more pipelines to Europe and new pipelines to the rest of Asia is a major strategic challenge for both the exporters and importers concerned, and the governments of countries which the route must cross. The first major project to be carried out on any route may pre-empt the development of alternatives for a long time. There may be competition between pipeline exports from eastern Siberia to north-east Asia and LNG from the Middle East to the same market. The export of LNG from the Middle East to South Asia is already a growing business and sets a benchmark for proposals to pipe gas to India from Iran. Nigeria, Trinidad and new discoveries offshore West Africa will supply the North American and European markets by sea as LNG,[8] providing competition for long-distance pipeline supplies from western Siberia. It is possible that in the long term the LNG trade will be sufficiently diversified, and there will be sufficient long-distance pipelines to

[8] James T. Jensen, 'The Outlook for Remote Gas Supplies', *Middle East Economic Survey*, 10 January 2000.

major markets, for there to be rough integration of world short-term gas markets which at present are essentially linked only by the medium-term alternative of switching to oil.

Natural gas demand to 2020

Figure 5.1 shows that in the last two decades of the twentieth century world natural gas consumption increased by 80%. In 2000, its share of primary energy consumption roughly equals that of coal at around 25%. By 2020, the main projections (discussed in Chapter 2) point to a share of 30%. This implies that gas consumption will double over the next 20 years, compared to an 80% increase over the last 20 years. In the past, gas replaced other fossil fuels. In the future, about half the gain in market share will be to replace the declining share of nuclear energy as nuclear supply is projected to stabilize and fall. The net effect

Figure 5.1: Growth in the gas share of world energy consumption

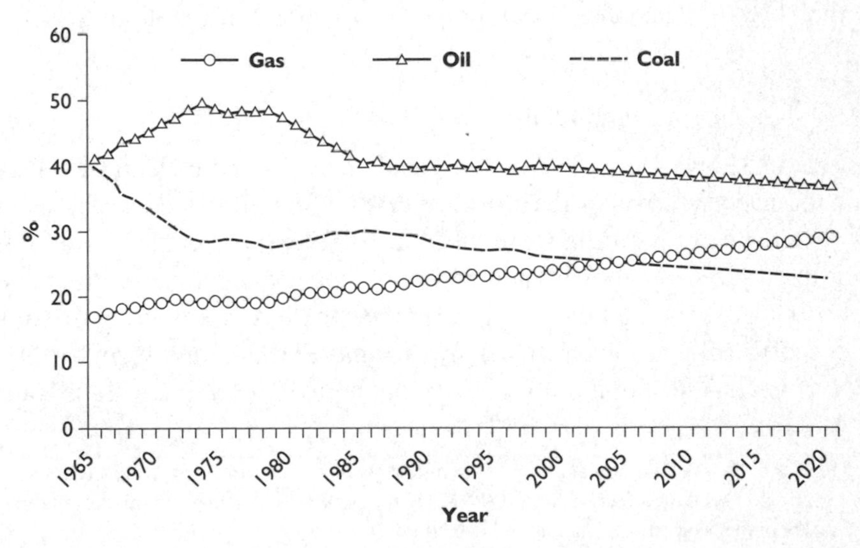

Note: Total primary energy in this graph is measured on an output basis. Calculation on a fossil fuel equivalent basis, as used in Chapter 2, would produce slightly lower market shares for all fossil fuels.

Source: BP Amoco *Statistical Review 2000.*

on the growth of GHG emissions will therefore be small. As described in Chapter 2, half the increase in gas demand is projected to take place in developing countries, compared with 45% of the increase over the past 20 years and their present share of about a quarter of total world gas consumption. For all regions, growth depends on increasing imports and is driven, as in the past, by various combinations of economic factors: availability of infrastructure to bring imports to the markets, the price of competing fuels, and the advantage accorded by environmental regulation to fuels low in sulphur and other pollutants. Strong policies on carbon emissions could reinforce the advantages of gas.

The opening of electricity and gas markets to internal and international competition is expected to continue, promoting aggressive competition and flexible pricing.

Different combinations of these factors will result in different patterns of evolution in different regions, and raise critical questions about the extent to which these projections may fail to be realized or may be exceeded. One common feature of almost all scenarios of growing gas consumption is the increasing importance of international and long-distance gas trade, which currently supplies about a quarter of world consumption.

Regional demand projections

Table 5.2 shows the projections of gas demand by the EIA and the IEA for the major regions of the world to 2020.[9] To compare these forecasts is a statistically hazardous exercise since the two organizations do not use the same geographical label to denote the same countries (the table uses the EIA regional categories).[10] Nevertheless, some comparisons are worthwhile, at least in general terms. Perhaps not surprisingly, both forecasts are similar up to 2010, but thereafter there are significant

[9] There is a more detailed breakdown in the industry consensus forecasts produced by the International Gas Union for the June 2000 World Gas Congress. Its broad totals fall within the range of the projections summarized in Chapter 2.

[10] Two major differences should be mentioned specifically: the EIA's 'North America' includes Mexico whereas the IEA's does not; and the EIA's figure for the 'former Soviet republics', which is what the IEA calls 'transition economies', includes a large number of countries in 'eastern Europe'.

divergences. World demand growth for the period up to 2020 is 0.5% per annum higher in the EIA forecast than in the IEA figures, which gives rise to a total demand more than 10% (or more than 400 mtoe) higher.

In the regional figures for 2020, one of the major differences between the two sets of estimates is in respect of North America. There the EIA is much more bullish – even taking Mexico out of the EIA figures, its figure for demand in 2020 is 834 mtoe, compared with the IEA's 676 mtoe. North America accounts for more than half of the difference in the world totals. The European and former Soviet figures are comparable, given the different geographical categorization. In Asia, the figures on the industrialized countries are similar but the EIA is significantly more optimistic about developing country demand, a subject to which we return below. For Africa, the IEA growth rates are double those of the EIA, but the overall numbers are too small to impact on the global total. In Latin America, the EIA is significantly more optimistic about demand expansion, but the difference does not amount to more than 50 mtoe in 2020. Middle East estimates are comparable, although the EIA sees most of the growth prior to 2010 and the IEA in the second decade of the century.

Table 5.2: Natural gas-demand scenarios by region, 1997–2020

	Projected demand (mtoe)					Annual growth rate (%)	
Region	1997	2010		2020		EIA	IEA
		IEA	EIA	IEA	EIA	1997–2020	1995–2020
North America	608	705	758	676	883	1.6	0.6
Europe	375		601		818		
Western	315	506	464	625	604	2.9	3.0
Eastern	61		138		214	5.6	n/a
Former Soviet Republics	459	647	587	835	746	2.1	2.1
Asia	207	445	445	663	728		
Industrialized	75	118	103	132	121	2.0	2.4
Developing	133	326	343	531	613	6.9	5.5–6.4
Africa	42	71	56	102	60	1.8	3.9
Latin America	68	185	198	306	357	7.5	4.9
Middle East	140	164	228	261	280	3.0	3.7
World	1,902	2,721	2,874	3,468	3,881	3.1	2.6

Sources: IEA: *World Energy Outlook 1998*, table 8.1, p. 134; EIA: *International Energy Outlook 2000*, table A5, p. 176.

Tables 5.3 and 5.4 show 'conventional vision' projections for gas demand in power generation and other 'stationary sectors'[11] over the period 1995–2020. These tables capture some of the detail in the different rates of increase in different regions, and the timing of these developments over the next two decades, which does not appear in Table 5.2 or the comparisons with oil in Chapter 2.

For *power generation*, the demand increases are extremely strong throughout the period, but especially in the OECD region up to 2010 and outside the OECD thereafter. For OECD countries, gas used for power generation nearly doubles in North America in 2010 compared with 1995, but thereafter the increase is very small. The picture is similar in the OECD Pacific region, while in Europe growth is strong throughout the period. In the non-OECD region, the biggest numerical increases (although not necessarily the highest rates of growth) are in the transition economies, East Asia, Latin America and the Middle East.

Table 5.3: Power generation gas-demand scenarios

Region	Projected demand (mtoe)				Annual growth rate, 1995–2000 (%)
	1995	Increase 1995–2010	Additional increase 2010–20	2020	
OECD	227.4	297.8	152.9	678.1	4.5
North America	131.6	121.9	17.2	270.8	2.9
Europe	55.0	139.4	124.5	318.9	7.3
Pacific	40.8	36.5	11.2	88.5	3.1
Non-OECD	337.2	171.4	281.3	798.9	3.5
Transition economies	228.0	39.1	83.7	350.8	1.7
Africa	15.3	10.2	19.3	52.9	5.1
China	0.7	11.7	11.2	23.6	14.9
East Asia	27.1	37.7	44.1	108.9	5.7
South Asia	11.8	14.6	27.7	54.1	6.3
Latin America	20.7	28.0	63.7	112.4	7.0
Middle East	33.5	22.1	40.6	96.2	4.3
World	564.6	469.2	443.2	1477.0	3.9

Source: IEA, *World Energy Outlook 1998*, table 8.4, p. 126.

[11] The IEA defines 'stationary sector' as the sum of non-electricity energy consumption in the industry, residential, commercial, public service, agricultural, other sectors, non-specified and non-energy sectors (*World Energy Outlook 1998*, pp. 466–7).

For the *stationary sector* the picture is rather different. Up to 2010 there is a 10% increase in OECD demand compared with 1995, but this already masks a substantial increase in European and (in terms of growth rates) Pacific demand, compared with a slight decline in North America. After 2010 the decline accelerates in North America and European demand also falls, with only a small increase being registered in the Pacific. The non-OECD region records strong, positive and fairly consistent growth throughout the period in each region. The regions with the strongest growth are the same as for power generation with the addition of South Asia.

Table 5.4: Stationary sector gas-demand scenarios

Region	Projected demand (mtoe)				Annual growth rate (%)
	1995	Increase 1995–2010	Increase 2010–20	2020	
OECD	612.8	60.8	–50.3	623.3	0.1
North America	357.6	–3.5	–46.0	308.1	–0.6
Europe	225.1	55.0	–6.2	273.9	0.8
Pacific	30.1	9.3	1.8	41.2	1.3
Non-OECD	369.2	253.7	245.1	866.2	3.5
Transition economies	219.1	91.9	87.7	398.7	2.4
Africa	10.8	6.2	6.0	23.0	3.1
China	13.0	23.1	10.5	46.6	5.2
East Asia	19.2	25.8	26.4	71.4	5.4
South Asia	18.9	35.9	37.0	91.8	6.5
Latin America	48.8	43.6	38.8	131.2	3.9
Middle East	39.4	25.4	38.8	103.6	3.9
World	982	312.7	203.8	1,498.5	1.7

Source: IEA, *World Energy Outlook 1998*, table 8.3, p. 125.

The new balance of gas markets

The picture suggests that, in global terms, the future demand for gas will be different from that of the past. Much of the industry was built on the requirements of a heating load. Up to 1980, more than 85% of global gas demand growth was centred on northern hemisphere industrialized countries with a heating requirement – North America (United States and Canada), the Soviet Union and Europe. By 1998 the domination of the northern hemisphere had fallen to 74% and the emphasis had begun to fall on power generation.[12] The projections in Table 5.2 suggest that by 2020, the northern hemisphere share will be not much above 50%. Tables 5.3 and 5.4 suggest that future demand will be increasingly based on electricity generation.

Natural gas prices

The IEA's 1998 'business as usual' gas prices for the three major OECD regions are shown in Table 5.5. European import prices are assumed flat for the period 1998–2010;[13] they rise by about \$1.30/mm Btu (40%) between 2010 and 2015 and remain flat around \$4.30 until 2020. Japanese LNG import prices are also almost flat from 1998 to 2010 and then increase by \$2/mm Btu or 50% for the period 2015–20. US wellhead prices are assumed flat or falling for the period 1998–2005; between 2005 and 2015 they more than double from \$2.13 to \$4.38/mm Btu, where they remain flat until 2020.[14] In relation to other fuels, the EU import price remains marginally below that of oil, and well above OECD steam coal import prices, throughout the period. The US wellhead price remains at parity with steam coal imports, and half that of crude oil imports, until its major increase starting in 2005. By 2010 US wellhead prices are double those of coal (not shown in the table). They remain significantly below crude oil throughout the period. Japanese LNG import prices remain in the same relationship to crude oil import prices throughout the period.

Price scenarios from the EU's 'Shared Analysis' project (POLES) are shown in Table 5.6.[15] In comparison with the IEA scenarios, the American price increases are much steeper for the period up to 2010 but moderate thereafter, the overall increase during the period being less dramatic than that projected by the IEA. However, the European price scenario is more dramatic than the IEA's because of the decline in

[12] Cedigaz, *Natural Gas in the World, 1999 Survey*, table 33, p. 72.

[13] Prices in this paragraph are given in 1999\$.

[14] IEA, *World Energy Outlook 1998*, table 2.4, p. 34.

[15] *The Shared Analysis Project, Economic Foundations for Energy Policy*, (henceforth POLES), prepared for the European Commission Directorate General for Energy, vol. 2: *World Energy Scenarios*, p. 25.

prices between 1990 and 2000. The scenario is for prices to rise by 50% by 2010 and nearly double over the 20-year period. Much less rapid shifts are foreseen for Asian gas prices, which increase by only 35% over the entire period.

Table 5.5: IEA assumptions for 'business as usual' gas prices

Price (1999$/mm Btu)	1997	1998–2010	2015–20
US natural gas wellhead price	2.45	2.13[a]	4.38
Natural gas import price into Europe	2.80	2.99	4.29
Japan LNG import price	3.83	4.03	6.04
IEA crude oil import price	3.45	3.62	5.34

[a] 1998–2005.

Source: IEA, *World Energy Outlook 1998*, table 2.4, p. 34, converted to 1999$ and mm Btu. Because of the successive conversions, numbers in this and some subsequent tables should be regarded as approximations.

Table 5.6: POLES assumptions for 'reference case' gas prices

Prices (1999$/mm Btu)	1990	2000	2010	2020
Gas import prices				
American market	2.31	2.39	3.80	3.91
European market	3.27	2.13	3.24	4.23
Asian market	4.41	3.71	4.67	5.00
World crude oil price	5.08	2.37	3.62	4.12

Source: European Commission Directorate General for Energy, *The Shared Analysis Project: EU Energy Outlook to 2020* (Brussels: European Commission, 1999), table 2.5, p. 23.

The reasoning behind the POLES scenarios is as follows:

> Separated from the North American and Asian gas markets, the Western European market shows a price decrease at the end of the nineties for local reasons: the production over-capacities, the beginning of the market liberalization and (as elsewhere) the low oil price. On the longer run, limitation of regional resources and the pressure of increasing demand, in particular in power generation, will make necessary the importation of gas from a number of distant fields from Russia, North Africa and the Middle East. This will require a slow increase of the gas price which is necessary for the financing of these projects. In these scenarios, the gas import price in Western Europe

will reach only $3.24/mm Btu in 2010 and $4.23/mm Btu in 2020. So it should limit the rapid development of new capital-intensive projects as will be the case of the new risks created by liberalization of national markets.

The EIA projects natural gas prices only in the United States and its scenarios are shown in Table 5.7. What is different about the EIA projections is that they cover both wellhead *and* end-user prices. The EIA increase in wellhead prices is in line with both IEA and POLES. But the end-user price projections are particularly interesting, showing:

- a decline in residential prices;
- relatively flat commercial gas prices; and
- a 32% increase in the industrial gas price.

This suggests that the introduction of liberalization in residential and commercial markets will exert downward pressure on prices in these sectors. By contrast, in the industrial sector, which was liberalized more than a decade ago, prices are projected to rise significantly.

Table 5.7: EIA gas-price projections

Average price, lower 48 states (1999$/mm Btu)	1998	2000	2010	2020
Wellhead prices	2.04	2.25	2.72	3.97
End-user prices:				
Residential	7.09	7.15	7.10	6.90
Commercial	5.66	5.99	5.99	5.93
Industrial	2.86	3.15	3.53	3.77

Source: EIA, *United States Energy Outlook 2000*, tables A1 and A3, converted to 1999$ and mm Btu (see source note to Table 5.5 above).

One of the most interesting aspects of Table 5.7 is that it allows some judgements to be made as to how the value of gas with respect to sales to different groups of end-users in the chain is projected to change through 2020. Unless those involved in the transportation and marketing of gas can significantly reduce costs, their margins – certainly in terms of sales to residential and commercial customers – seem bound to fall during this period, whereas margins on sales to industrial customers seem likely to increase. In North America, therefore, on the basis of this evidence, the big margin 'winners' seem likely to be producers. As we shall see, this is not the case elsewhere in the world, but the data on which these judgements must be based are far less robust.

Market overview by region

A brief survey of recent developments in regional gas markets will help to highlight the critical issues determining the growth in gas markets. The high transport costs of gas, relative to other fuels, have meant that in most regions high and growing gas shares in the energy mix have been associated with the presence of an indigenous gas supply and the absence of alternatives such as low-cost domestic coal or nuclear power.

North America

The North American market, unlike markets in the other regions discussed below, is not 'national', it is continental. From the Liard Basin in the Northwest Territories of Canada in the north, to Mexico in the south, gas competes with gas.[16] It is a fully fungible market with no or few barriers to entry for sellers or buyers, instantaneous price discovery and market mechanisms to manage risk.

The United States is the largest consuming and the second largest producing country in the world, using 27% of the total global consumption. Natural gas's share of primary energy fell from around 33% in the early 1970s to around 25% in the mid-1980s, prior to the deregulation of the industry. Since deregulation, market share has increased slowly to 27%. The United States is also the largest single importing country, with imports in 1998 equivalent to about 60% of imports to Europe as a whole and 36% more than imports to Japan. The volume of imports (mainly from Canada) has increased nearly fourfold since 1980. Natural gas consumption is projected to expand to one-and-a-half times the present level by 2015. The major reasons for the increase are environmental protection and the expected decline in nuclear power

[16] The EIA definition includes Mexico.

generation. There are major uncertainties about the capacity of the United States to expand gas production, which has been roughly unchanged for the past five years. The sudden 'spike' in US spot and futures prices in 2000 raised wellhead prices to levels not seen since the early 1980s. Imports are expected to increase by means of additional pipeline flows from Canada and an expansion of LNG imports (from the Caribbean, South America and West Africa).

The process of 'deregulation' – the removal of price controls and market monopolies in both electricity and gas – led to a fundamental restructuring of the US gas industry, with a new emphasis on trading, distribution and marketing at regional or national rather than at state level. Gas prices throughout the region are based on those generated in the futures market – principally by the New York Mercantile Exchange (NYMEX) – with 'basis' differentials for transportation quoted at market hubs. Companies have merged and demerged. Open access to the regulated pipelines as far as to final consumers means that consumers can buy gas from anyone, at any time, if willing sellers can be found. Gas trading operations have been separated from transportation and distribution. The business models of the players have changed rapidly and are almost beyond comparison with those of 15 years previously.

Europe

Natural gas consumption has been growing steadily in Europe,[17] which uses about 20% of the world's gas. New environmental regulations, lack of growth of nuclear power, and a decrease in indigenous coal production are the main reasons for the increase. The IEA and EU have reversed policies put in place in the 1970s to restrict or prevent the building of new gas-fired power stations. Technical improvements in the efficiency of combined cycle gas turbine power generators mean that during the 1990s, natural gas had significant economic advantages for power generation, including high efficiency, rapid development of plant and low capital cost.

[17] For Europe, see Jonathan P. Stern, *Competition and Liberalisation in European Gas Markets: A Diversity of Models* (London and New York: RIIA/Brookings Institution, 1998).

New business models for European gas are also emerging. In the UK, liberalization and privatization of the gas and electricity markets enabled gas-fired power plants to be developed rapidly, and to achieve lower overall generation costs, although there was a pause in government licensing of new gas-fired generation plants in the UK during a crisis in the coal industry in 1999. The UK gas market is fully liberalized, with prices determined by gas-to-gas competition and an active spot and futures market. The 'interconnector' gas pipeline from Britain to the European continent, completed in 1998, has created a connection between the British and other north-west European physical and commercial markets for short-term gas supply.

Continental Europe is becoming a market dominated by imports (currently 40% of consumption). Taken together, Europe's main natural gas sources (the UK, Norway, the Netherlands, Algeria and Russia) have sufficient gas reserves to support several decades of European demand. There is a well-developed infrastructure to bring gas from these countries to Europe, and currently an over-supply of both gas and capacity. However, the extent to which the present surplus can carry Europe through to 2010 or 2020 (as opposed to 2005) is uncertain. It is possible that additional supply sources and infrastructure corridors will be needed from further afield, such as the Caspian region (the Caucasus countries and Central Asia) and the Middle East.

Continental Europe is in the process of liberalization initiated by the European Gas Directive of 1999. The first level of market opening was initiated in August 2000 and in the majority of countries is moving faster than foreseen by the Directive. How fast this will change the present pricing structure of long-term take-or-pay contracts, indexed (largely) to oil product prices, towards gas-to-gas competition with prices set by spot and futures markets, is a matter of considerable debate. Competition and pricing developments in European electricity markets have moved rapidly in this direction since the opening of the markets in 1999. The development of continental European energy exchanges to take advantage of increasingly commoditized energy markets suggests gas will follow this trend. The gas industry will also have to develop new business models to deal with more short-term trading, more gas-to-gas competition, and more competitors physically linked to the grid to which the directive provides access.

Transition economies

In the countries of the FSU natural gas is well established as a source of energy, accounting for around 50% of primary energy demand in Russia, Ukraine and Central Asian countries.[18] The break-up of the Soviet Union and subsequent transition of those countries to market-based economic systems has given rise to severe economic dislocation with a major impact on energy and natural gas demand. While natural gas production has declined somewhat since 1991, it has been the least affected of all the fossil fuels. Demand has also declined but did not reflect the enormous fall in industrial production and GDP (more than 50%) which took place during the 1990s. A major reason for this is the phenomenon of 'non-payment' whereby consumers of all kinds continue to receive gas despite not paying for it in cash, on time, or (sometimes) at all. In the late 1990s, less than 20% of gas deliveries to domestic customers by Russia's Gazprom were paid promptly and in cash. Unless payment is more strictly enforced, it will be impossible to make new investments in production or transmission to serve the domestic economy. However, as and when payment is more strictly enforced one would expect to see demand fall as market signals generate behavioural change and new investments in energy saving and efficiency. In such a situation, it is not obvious that gas demand in this region will increase strongly (or perhaps at all) over the next two decades.

---

[18] For the transition economies, see Jonathan P. Stern, 'Soviet and Russian Gas: the Origins and Evolution of Gazprom's Export Strategy', in Robert Mabro and Ian Wybrew-Bond, eds, *Gas to Europe: The Strategies of the Four Major Suppliers* (Oxford: Oxford University Press, 1999), pp. 135–200.

Asia

Asia accounted for only 12% of the world's total natural gas consumption in 1999.[19] The lack of a natural gas infrastructure is generally thought to be the reason for the low gas consumption in countries without domestic gas supplies (unlike Indonesia and Malaysia). LNG importers – Japan, Korea and Taiwan – have increased their consumption rapidly, but growth slowed down in the 1990s, even before the crisis in the latter part of the decade. The biggest constraint on the increase in natural gas use may be the high price of LNG. There is a fundamental trade gap between the energy-importing areas of South and East Asia and the gas-exporting areas of West Asia (the Middle East) and Asian Russia.[20] To make this trade possible requires, first, substantial new investment in long-distance, cross-border transportation (either by pipeline or LNG) from Russia or the Middle East, and second, investment in local distribution, particularly in countries without major gas infrastructure. To provide imports of gas at competitive prices in such conditions is a major challenge both for the governments of the countries concerned and for the international gas industry. The outcome will affect the gas share of world energy consumption, because of the size and expected growth of Asian energy consumption. Figure 5.2 compares the projected growth in the 'conventional vision' in different regions with the current levels of consumption and market shares.

In the industrial sectors of most Asian markets, natural gas with a high delivered price and a high transmission cost cannot compete with other fuels, especially cheaper coal. In the power sector also, the high price of LNG, combined with the attractive economics of coal use, limit growth prospects for gas use. LNG import prices may be forced down under the twin impact of competition among suppliers and liberalization and competition in import markets. The Asian economic

---

[19] See James T. Jensen, 'International Gas Markets, Supplies and Trade', presentation to 22nd Oxford Energy Seminar, 30 Aug. 2000 (Wellesley, USA: Jensen Associates Inc., 2000).

[20] See John Mitchell and Christiaan Vrolijk, *Bridging Asia's Energy Gaps*, RIIA Briefing Paper no. 41 (London: Royal Institute of International Affairs, 1998).

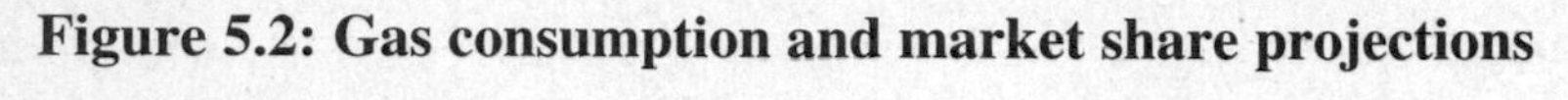

Figure 5.2: Gas consumption and market share projections

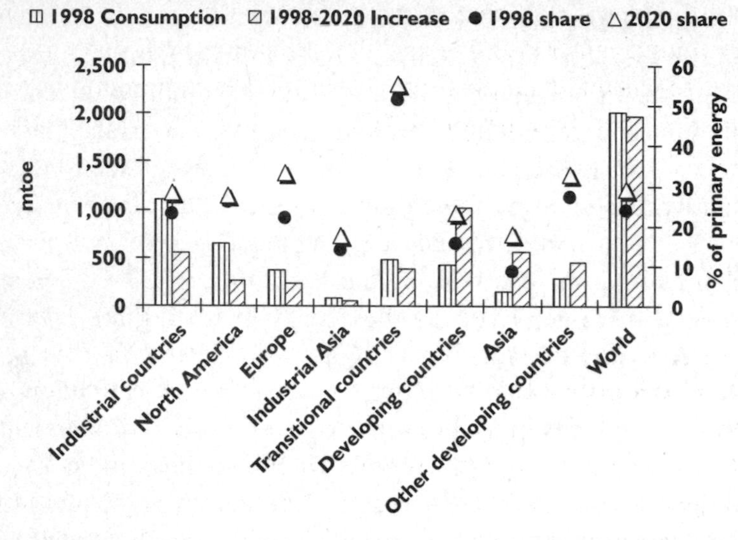

crisis of the late 1990s and subsequent recession in countries such as Japan made clear the need for lower LNG import prices. The liberalization policy of the Japanese government in particular is aimed at reducing the cost of natural gas and electricity derived to Japanese industrial consumers.

Natural gas resources are abundant in South-east Asia as a region, but not in every country: Thailand is a long-term gas importer and Indonesia is a long-term gas exporter. There are major pipeline developments under way in the region.[21] Consumption of natural gas in parts of the region has been growing faster than that of other fuels. Since 1990 the annual growth rate for gas for the region as a whole has been 10.6%. In this region, as elsewhere in the world, power utilities have been privatized and the responsibility for generation is being shifted to the private sector or independent power producers (IPPs). Problems may continue to arise when new projects look for electricity prices that exceed the prices which formerly prevailed under price controls and with subsidized state enterprises. Domestic and foreign IPPs have tended to use natural gas in new stations as gas-fired generation is commercially more attractive than other fuels, especially coal, and is better equipped

[21] Jensen, 'Natural Gas Policy Issues for the Asian Region'.

to meet any environmental standards that may be introduced. Natural gas utilization will continue to be promoted even though there may be some difficulty in financing such schemes. However, delivering LNG or pipeline gas into a competitive power market comprising large numbers of IPPs will be a major challenge, and one which remains to be faced by developers.

China and India

Events in China and India will have a considerable impact on natural gas demand in Asia over the next two decades.[22] Table 5.8 shows EIA and IEA projections for Chinese and Indian gas demand. Environmental problems loom large in both countries but particularly in China, where improving urban air quality is a key driver of policy favouring increased gas usage. In order to develop a pipeline network to supply gas to major cities, a feasibility study on a huge national gas infrastructure has been undertaken. The plan would be to augment domestic production with import trunk lines from Russia in the north and from Central Asia in the west, designed to ensure security of supply.[23] On the country's east coast a number of proposed LNG projects – one of which has already been approved – would bring gas to these regions of the country. Currently, natural gas forms just 2% of China's primary energy, but these ambitious plans would increase this figure to 6–7% by 2020. Whether or not such plans are sufficiently viable to attract private sector funding, and whether the Chinese government will provide financial support if they are not, remains to be seen.

Gas constitutes 11% of India's energy mix, but a decline in domestic production is imminent. While imports of pipeline gas from Central Asian (Turkmenistan), Middle Eastern (Iran and the Gulf) and South-east

[22] For India, see EIA, *International Energy Outlook 2000*, p. 57; IEA, *World Energy Outlook 1998*, pp. 323–98. For China, see IEA, *China's Quest for Energy Security* (Paris: OECD, 2000); Lan Quan and Keun-Wook Paik, *China Natural Gas Report* (Beijing: Xinhua News Agency; London: Royal Institute of International Affairs, 1998).

[23] See Keun-Wook Paik, *Gas and Oil in Northeast Asia* (London: Royal Institute of International Affairs/Earthscan, 1995).

Asian (Burma) countries, as well as Bangladesh, would be commercially feasible, progress is hampered by transit and security issues associated with neighbouring countries, in particular, Pakistan and Bangladesh.[24] In the absence of solutions to these issues, all India's gas imports will be in the form of LNG.

Table 5.8: Chinese and Indian gas-demand projections (mtoe)

	1997	2010		2020 (annual % increase, 1997–2000)	
		EIA	IEA	EIA	IEA
China	16.8	93.6	56.6	206.4 (11.2%)	80.7 (6.5%)
India	19.2	64.8		115.2 (7.9%)	
Total	36.0	158.4		321.6	

Sources: IEA, *World Energy Outlook 1998*, table 8.1, p. 128; EIA, *International Energy Outlook 2000*, table A5, p. 176.

For both China and India, the constraints are the gas price and the infrastructure. It is difficult to see who will invest in the infrastructure, including power plants, if the price of natural gas must be lowered to a level at which gas-fired power plants can be cost competitive relative to domestic coal. Even in the non-power-generation sector, price levels would need to be raised in order to make projects commercially viable.

The outlook for gas in China and India is extremely uncertain. Potential demand is huge, but is hampered by a shortage of conveniently located supply and low (government-controlled) prices to end-consumers. Discovery of substantial indigenous gas reserves in either of these two countries could make a major difference to the prospects for gas demand over the next two decades. These prospects in turn will have a significant impact on the demand for other fuels – coal and oil – which are likely to increase enormously over the same period. What Table 5.8 shows is that if demand increases as projected by the EIA, this would make a very substantial addition (nearly 300 mtoe) to global gas demand in 2020.

[24] Ardhendu Sen, 'Natural Gas Imports into South Asia: A Study in International Relations', *Energy Policy*, vol. 28 (2000), pp. 763–70.

Latin America

In Latin America the consumption of natural gas is increasing rapidly. Venezuela and Argentina are the largest consumers, using indigenous gas, followed by Brazil, Colombia and Chile.[25] The construction in this region of several long-distance pipelines and the development plans to expand intra- and international natural gas trade are now attracting the attention of the multinational energy companies. For example, the Bolivia–Brazil pipeline project, which is more than 3,100 km long and capable of transmitting 10bn cu metres per year, was completed and started operation in 1998. Privatization of, and foreign investment in, national utility companies has been a major driving force for gas development over the 1990s. The rapid development of gas markets in this decade serves as an example of how this may be achieved in other regions.[26]

Africa

The twin foci of gas development are the countries of the Maghreb in the north and Nigeria in sub-Saharan Africa. In North Africa, Algeria is already among the world's largest gas-exporting countries, and there are projects under way in Libya and Egypt (see below) which would further expand gas use and exports from this part of the continent. In sub-Saharan Africa the fuel is under-represented in energy balances. Given the size of the African continent and the magnitude of the available resource base, demand is extremely small. Fulfilling the increases anticipated (see Table 5.2) would require projects such as the West African Gas Pipeline and the Mozambique–South Africa pipeline to go ahead.

---

[25] For Latin America, see James T. Jensen, *The Development of International Gas Transportation Systems* – Latin America in a World Context, Institute of Gas Technology, 21 June 1999.

[26] For a detailed review of current developments, see also 'South America: A Flair for Gas', *Petroleum Economist*, May 2000, pp. 30–4.

Country examples

The drive for gas

The aim of this section is to examine four countries where gas consumption has expanded greatly during the 1990s: Korea, Brazil, Turkey and Egypt. Of these, Egypt may be regarded as a long-term potential exporter, the others as long-term gas importers. All four countries started to import or use natural gas on a large scale in the latter half of the 1980s. At that time dependency on oil imports was already large and expected to increase. If these are the similarities between the countries, there are also important differences in the situations in which they found themselves. Table 5.9 presents basic data on the four countries.

Table 5.9: Four countries going for gas

	Brazil		Egypt		Korea		Turkey	
	1988	*1998*	*1988*	*1998*	*1988*	*1998*	*1988*	*1998*
Population (m)	143	166	50	61	42	46	54	63
GNP/capita (US$)	2270	4630	760	1290	4110	8600	1570	3160
Energy consumption (mtoe)	89.5	126.1	29.0	40.3	74.2	167.1	48.4	62.8
Share of TPES (%)								
Oil	65.8	66.0	75.9	67.7	48.0	55.8	46.1	47.9
Gas	3.8	4.6	21.0	27.3	4.0	8.4	6.2	14.2
Coal	10.6	8.8	2.8	2.5	32.9	21.6	46.1	32.2
Nuclear	0.7	0.6	0.0	0.0	18.3	13.8	0.0	0.0
Hydro	19.9	19.8	2.8	2.7	0.7	0.3	4.1	5.7
	1989	*1998*	*1989*	*1998*	*1989*	*1998*	*1989*	*1998*
Oil production	0.8	1.2	0.9	0.9	0	0	0.06	0.07
Net oil imports (mb/d)	n/a	0.7	n/a	–0.4	n/a	2.0	n/a	0.6
	1989	*1998*	*1989*	*1998*	*1989*	*1998*	*1989*	*1998*
Coal production	7.4	5.2	0	0.4	22.9	4.8	57.6	67.5
Net coal imports (m US short tons)	n/a	15.4	n/a	1.3	n/a	57.7	n/a	7.4
	1988 end	*1998 end*	*1988 end*	*1998 end*	*1988 end*	*1998 end*	*1988 end*	*1998 end*
Natural gas reserves (trn cu metres)	0.1	0.23	0.32	0.89	0	0	n/a	n/a

In the ten years between 1988 and 1998, natural gas consumption in the four countries increased rapidly, almost doubling in Brazil, more than doubling in Egypt, expanding by five times in Korea and by nine times in Turkey. The share of natural gas in the primary energy mix increased, and it contributed to the diversity (and hence security) of energy supply in each country. Table 5.10 summarizes the position in the four countries

Table 5.10: The growth of natural gas in four countries, 1988–98

	Increase in gas demand (bn cu metres)		Gas as % of primary energy demand	
	1988	1998	1988	1998
Egypt	5.9	12.2	21.0	27.3
Korea	3.0	15.6	4.0	8.4
Brazil	3.7	6.5	3.8	4.6
Turkey	1.1	9.9	6.2	14.2

One of the most important reasons for the introduction and expansion of natural gas was the need for additional energy supply to meet an anticipated growth in demand, especially for power generation. The reasons for the choice of natural gas varied among the countries. Only in Egypt were local resources a major factor. All four had been seeking to reduce oil imports. Korea and Turkey faced declining domestic coal availability and rising urban pollution problems aggravated by household coal burning, as well as power station use of coal. Turkey was, and is, advantageously placed to solicit competing supplies from Russia, the Transcaucasus and Central Asia (as well as LNG from Algeria and potentially pipeline gas from Egypt).

Energy resources

Brazil has expanded oil production to cover about 50% of its consumption, but gas discoveries have been limited. This situation may change as a result of the removal of the Petrobras monopoly upstream in 1998, and the further development of technology for deeper water exploration and production (in which Petrobras is a world leader). Brazil has therefore committed itself to international agreements and to internal opening of the gas market to foreign investment for imports from Bolivia.

Egypt is an oil exporter as well as an oil market. Its ratio of oil reserves to production is around 10:1, in contrast to a ratio of gas reserves to production of around 70:1. Egyptian policy is therefore placing more emphasis on developing gas consumption as well as investigating the possibilities of exporting gas to countries in the region.[27] In the decade 1988–98, Egypt was the only one of the four cases where there was a significant substitution of gas for oil in the domestic market. Turkey and Korea have no significant domestic gas resources.

In other countries the history is one of gas share increasing at the expense of coal. The future may be different. Korea and Turkey, lacking oil and gas resources, had other reasons for the introduction of natural gas. Energy consumption had been increasing rapidly in the early 1980s, a trend which was expected to continue. In fact, it expanded 2.3 times in Korea between 1988 and 1998. Dependence on oil, almost all of which is imported, was already considerable. Production of coal, which is the only significant indigenous energy source in either country, had peaked in Turkey and was declining in Korea. Also, coal was causing serious environmental problems, particularly in Korea. These conditions led to the introduction of natural gas, principally on the initiative of government because of the enormous capital investment required for infrastructure development which would not have been attractive to the private sector. In order to promote the import of LNG and pipeline gas, the national corporations of Kogas and Botas were established by the Korean and Turkish governments respectively in the 1980s.

Availability of imports

Geographical proximity to resources and the influence of neighbouring countries were also contributing factors to the introduction of gas in

[27] For a recent review, see 'Egypt: Gas Explosion', *Petroleum Economist*, September 1999, esp. p. 62

these countries. Turkey is surrounded by countries with large exportable gas surpluses to the north, south and east, and its neighbours import gas from Russia. Japan, which is the nearest country to Korea, started to import LNG at the end of the 1960s and was expanding consumption rapidly in the 1980s. These facts may have been important influences on Korea and Turkey. For Korea, LNG was the only option, but the possibility of imports from eastern Siberia became part of the long-term supply framework during the 1990s.[28]

Government policy

Two different patterns of development of gas use can be identified in these four countries: government-led development in Korea and Turkey, and private-led development in Egypt and Brazil. In the latter countries liberalization (including the removal of energy price controls) and the privatization of national companies gave rise to an influx of foreign investment to expand the market for natural gas.

Both the Egyptian and the Brazilian gas industries started from government-led development, but the results were unsatisfactory, and the governments decided to introduce private (including foreign) capital into their industries. In Brazil, the greatest impact in the 1990s was in the downstream (transmission and distribution) sector, while in Egypt foreign investment concentrated in the upstream sector.

Table 5.11: Natural gas-demand prospects in Turkey, by sector (bn cu meter)

	Residential	Industry	Power generation	Fertilizer
1998	2.7	1.6	5.5	0.5
2000	3.4	3.4	12.3	0.8
2005	6.6	8.8	29.2	0.9
2010	8.4	11.0	34.2	0.9

Source: Cedigaz, *Natural Gas in the World*, 1998 survey (Botas).

[28] See Keun-Wook Paik and Jae-Tong Choi, *Pipeline Gas in North-East Asia*, Briefing Paper no. 39 (London: Royal Institute of International Affairs, 1998).

Brazil

In Brazil, natural gas infrastructure developed considerably during the 1990s. After a decade of negotiation, the first stage of the Bolivia–Brazil pipeline project started operation in 1999 and additional pipelines are under discussion. To raise capital, state governments have begun to sell their state natural gas distributing companies to the private sector. Privatization plans are under way and are changing the structure of the energy sector, especially natural gas. As a result, international capital markets are becoming easier to access, even though domestic competition is increasingly encouraged. Foreign companies like British Gas, Shell and Enron have already invested in local gas distributors. Owing to the determination of state governments to privatize their gas distributing companies, British Gas and Shell gained a 62% share of Comgas in São Paulo. There are plans to construct further natural gas pipelines linking Brazil to neighbouring countries.

Egypt

Foreign companies (British Gas, BP Amoco, ENI-Agip and Shell) are involved in exploration and production. An LNG export project is under way and future LNG and pipeline projects are in the planning stage. The Egyptian government plans to accelerate its programme for the privatization of state-owned enterprises, which would bring foreign capital into the downstream gas and electricity industries. Natural gas is consumed at the thermal power plants, which are owned by the Egyptian Electricity Authority (EEA). EEA is the major candidate for privatization and future gas demand is expected to expand by private sector or build–own–operate–transfer (BOOT) projects. Domestic gas consumers are to be served by several private gas distributors working through franchises which were awarded with the participation of foreign investors. Around 20,000 taxis have already been converted to use CNG. Natural gas is produced and consumed by private initiatives.

Korea

In Korea, where all supply depends on import, a national corporation (Kogas) has been established by the government with a monopoly on imports of LNG. Kogas is also the sole operator permitted to sell gas to the power utility (Kepco) and local city gas companies. The natural gas infrastructure (LNG receiving terminals and a trunk-line system of 2,300 km which circles the peninsula) were developed by Kogas. All the capital investment was provided by the government. Natural gas supply started in Seoul in 1986 with a 'big bang' of economic growth called a 'miracle of Hangang' (the name of a river which flows through Seoul). At first natural gas was consumed mainly in power stations, which could be constructed with relatively short lead times, to supply electricity to the rapidly increasing population south of Seoul. After that, consumption by city gas companies increased at a high rate, and with the extension of the trunk-line system from the LNG receiving terminal near Seoul City, gas consumption has finally overtaken demand in the power sector.

Table 5.12: Natural gas consumption in Korea, by sector

	LNG consumption (000 tonnes)						Annual growth
	1993	1994	1995	1996	1997	1998	rate (%)
City gas							
Household	1,157	1,612	2,300	3,089	3,768	3,864	27.3
Industry	239	306	451	661	1,043	1,394	42.3
Others	451	536	666	832	959	974	16.6
Subtotal	1,847	2,454	3,417	4,582	5,770	6,232	27.5
Electricity	2,518	3,329	3,562	4,622	5,377	4,189	10.7
Total	4,365	5,783	6,979	9,204	11,147	10,421	19.0

Source: Kogas, *Gas Facts*, 1999.

Kepco, a national corporation in charge of generation, transmission and distribution of electricity, is reluctant to increase natural gas consumption. One of the major reasons is the high sales price of natural gas from Kogas, the only source. Distribution costs, including the depreciation cost of the total trunk-line system, is added to the LNG import price. Kepco is already in the middle of a liberalization process and is being urged to reduce its costs in order to survive. Current prospects for natural gas demand in the electricity sector are not optimistic.

Turkey

Consumption is projected to rise as part of Turkey's policy of substituting natural gas for high-sulphur fuel oil and coal. Gas production in Turkey is limited by lack of resources, despite the involvement of foreign companies. Most of Turkish gas demand is met by imports. In 1998, 65% of imports came from Russia via Europe and 35% was delivered as LNG from Algeria and Qatar. All the pipeline infrastructure and all import contracts are handled by the national corporation, Botas. As a government monopoly Botas is well placed to manage and encourage competition among suppliers for its large and growing market. Turkey faces a rich choice of pipeline suppliers and supply routes:

- Russian gas, delivered through Gazprom's lines via the European route and through the Blue Stream pipeline across the Black Sea;
- Iranian pipeline gas;
- Turkmen pipeline gas via Russia or across the Caspian Sea;
- Azeri pipeline gas.

The story of the former Soviet suppliers in Russia, Central Asia and the Caspian, and their various gas pipeline projects, is a classic example of the importance of gas transportation as a vehicle for competition among gas producers, as well as a vehicle for geopolitical rivalry.[29]

Turkey also has various potential sources of LNG imports, including Algeria, Nigeria, Trinidad and Egypt.

[29] For a many-faceted analysis, see William Ascher and Natalia Mirovitskaya, eds, *The Caspian Sea: A Quest for Environmental Security*, NATO Science Series 2, vol. 67 (Dordrecht: Kluwer, February 2000).

In Korea, the economy and the natural gas industry have now grown sufficiently that government assistance may no longer be necessary to raise capital. The Korean electricity industry is being liberalized and Kogas is in the process of being privatized. In Turkey, gas demand is expected to increase more than fivefold by 2010 (see Table 5.11). Meeting this demand will require enormous investment in gas and electricity industries, some (and perhaps a large) part of which will need to come from foreign sources.

Critical questions for gas and oil markets

To fulfil, or exceed, the 'conventional vision' of accelerating the historic shift to gas will require increasing use of gas in new geographical markets, in the conventional power sector worldwide, and in markets created by new technology in sectors such as transport and distributed power generation.[30] Unless there is a failure to invest among Middle East oil exporters over the next decade, there is unlikely to be the kind of price incentive for gas which was created by the oil prices of the period 1975–85. The agendas will be driven by some combination of competition, regulatory reform and technology. The final impact of gas on the oil markets depends on whether the kind of gas industry revolution which has taken place in North America is replicated in Europe and Japan and spreads to developing countries. Will technology reduce transportation costs and open new market sector for gas? Will a common pattern emerge for intergovernmental agreements to underpin cross-border gas infrastructure projects? How far will the resource owners have to carry transportation and market risks in order to get their reserves developed earlier than those of their competitors?

Unfinished agendas

Answering these questions requires solutions to one or more of a number of problems in specific situations:

[30] For a layperson's outline of distributed generation and the related technologies, see Ann Chambers, *Natural Gas, Electric Power* (Tulsa: Pennwell, 1999), ch. 13, pp. 166–79.

- Reducing the technical cost of building and operating pipelines or LNG systems for currently identified gas resources remote from the large established and projected markets.
- Investing in power stations and/or industrial use to achieve a rapid build-up of volume in countries where there is no existing heat load. Potential industrial demand in a rapidly developing economy is difficult to forecast and may itself depend on the delivery of cheap energy.[31] The technologies of micro-generation by fuel cells or small gas turbines in industrial heat and power plants may be a key to such investment.[32]
- Reducing the price of gas to end-users, especially in Asian markets where coal or coal-based electricity is an alternative.
- Finding new and innovative ways (for the gas industry) of managing the financial risk of large-scale infrastructure for new markets, especially where rates of growth are uncertain and there are no strong distribution companies.
- Building an international legal and institutional framework for gas (and electricity) in transit to, and through, multiple countries in which all parties – importers, consumers and investors – can have confidence.
- The influence of privatization and liberalization in all regions and countries, in both the electricity and gas sectors, which promises to change the traditional monopoly/dominant positions which national and regional utility companies have enjoyed.
- Integrating new gas supplies into the structure of gas and electricity markets that are changing rapidly in the direction of liberalization and increased competition.
- Balancing the lower pollution impacts of using gas with the environmental and social impacts of new gas supply projects in countries where governments are considered to have poor records in terms of the human and political rights of their populations, or alleged

[31] For a detailed example of the practical and theoretical problems involved in a project for creating industrial demand for gas in Indonesia, see Willem J. H. Groenendaal, *The Economic Appraisal of Natural Gas Projects* (Oxford: Oxford University Press, 1998).
[32] See e.g. 'The Dawn of Micropower', *The Economist*, 5 August 2000, pp. 99–101.

sponsorship of international terrorism. As described in Chapter 8, the plans of international investors will be challenged in countries where national governments and investors take an interest in such matters, and by multilateral agencies like the World Bank if these are involved.

The following sections look at some of these issues in more detail.

Gas resources and transportation to remote markets

Table 5.1 showed that proved gas reserves and resources are abundant and, in total, well able to cover anticipated increases in demand even in the absence of additional discoveries.[33] The exceptions to this general

Box 5.1: New technologies and gas markets

- Fuel cells for vehicles: the technology is largely proven today – vehicles may be cost-competitive with conventional vehicles given sufficient scale of manufacture. There is the prospect of significant market penetration in 2010–20, but this depends on customer acceptability and government regulation and incentives (taxation). Storage technology is the key. The infrastructure could also be under development within this timeframe. Hydrogen is a long-term fuel solution but, without massive expansion of nuclear power, depends on fossil fuel sources for its production.
- Urban transport: experience in California and elsewhere in the US has shown that government mandates to compel city fleet transport to use compressed natural gas can create a market for gas in this area. This solution – as an alternative to 'city diesel' – may continue in parallel with the development of fuel cell vehicles. (See RIIA Briefing Paper No. 9, *Oil for Wheels*.)
- Fuel cells for stationary uses: in Japan work is progressing towards commerciality of a fuel cell co-generation system for domestic use in 2005. Sales could increase residential gas volumes significantly by 2010. While both vehicular and stationary fuel cells have a substantially positive effect on local pollution levels, their benefits in terms of reduced CO_2 emissions are more modest.
- Micro power generation (turbines): there will no doubt be increasing gas use in power generation, but the jury must remain out as to whether the likely move towards smaller generating units will actually increase gas use further.
- Gas-to-liquids: can they compete with oil in the light of relative gas and liquid product prices? Widespread commercial development is approaching, but requires feed gas at low net cost. It is important to emphasize the scale of gas stranded from markets (over 250 trillion cubic feet in over 2000 fields worldwide), the applicability of gas-to-liquids to a wide range of reserve sizes, and the cleanliness of the liquids produced.

[33] James T. Jensen, 'The Outlook for Remote Gas Supplies', *Middle East Economic Survey*, 10 January 2000.

rule are in China and India, where the discovery of substantial fields could transform the outlook for the fuel.

For a significant proportion of the global resource base, both domestic and export markets are distant, and transport costs are relatively high. This commercial problem is compounded by transit issues – the fact that pipeline routes must cross many other countries, including some which export in their own right. Issues of viability, security and international interdependence are critical. These factors may make technologies (e.g. gas-to-liquids) which avoid gas transport potentially attractive to resource owners.

Meanwhile, it remains the case that the owners of large resources will compete with one another and with oil. The ability to develop low-cost transportation will be the key to success. Early high throughput is critical for achieving low unit costs of transportation.

Gas infrastructure: technological, legal and financial issues

The development of the natural gas infrastructure depends on three key areas.

Technological advances are progressively enabling the installation and operation of gas pipelines in ever deeper water, offering bilateral national transportation solutions in regions where multilateral international agreement on transit arrangements appears elusive. Exploration and the production of gas at 1,000–1,500 metres in the Gulf of Mexico has already allowed the commercialization of reserves believed to be uneconomic only a few years ago. The Blue Stream project is being designed to take Russian gas directly to Turkey across the Black Sea, at water depths down to 2,150 metres. Technology has another important role: to reduce costs. Technological developments continue to bring down costs throughout the pipeline and LNG gas chains. BP Amoco, for example, has stated that the costs of the Trinidad LNG project will permit supply into the United States at prices that are competitive with pipeline supply (around $2/mmbtu).

As gas becomes a more widely available and traded fuel, so resources cross international boundaries more frequently. The lack of a widely accepted *international legal regime* for cross-border energy transfers may be a major factor inhibiting investment in infrastructure considered to be economically viable. This has been experienced particularly in respect of existing infrastructure in, and exports from, Russia and other FSU. However, the future development of cross-border networks in a variety of regions will benefit from an internationally accepted, enforceable legal regime. The Energy Charter Treaty signed by a number of European and Eurasian countries, and ratified by most of them, may provide a basis for such a regime in the future. The Energy Charter Treaty Conference has approved a model transit agreement designed to provide a framework for such investments.[34] However, the treaty and the model agreement embody features which some national governments may find difficult to accept: national treatment for foreign investment; a dispute settlement procedure which gives foreign (but not domestic) investors access to international arbitration against the government, and rights to demand government support for facilities to establish competing transport systems.[35]

Financing infrastructure will be particularly important in regions where gas resources are abundant. In South-east Asia, gas transmission, distribution and IPP projects have begun to take off in countries such as Malaysia, Thailand, Indonesia and the Philippines. Foreign investors are interested in participating in these ventures, but obstacles exist, particularly in newly liberalizing markets. There is often no agreement as to the appropriate balance of state/private and national/foreign finance, or on the role of international financial organizations. In Korea and Taiwan, the governments themselves have successfully constructed pipeline systems. But in South America and South-east Asia, the liberalization of the energy sector and the privatization of national companies preceded the introduction of foreign capital required to develop the gas infrastructure. It is not clear which models may be applicable in other parts of Asia, especially in China and India.

Although these developments widen the options for economic gas transportation, they leave unresolved the question of whether owners

---

[34] Energy Charter Secretariat, *Activity Report 2000*, <www.encharter.org>.

[35] See Bernard Mommer, *Oil Prices and Fiscal Regimes* (Oxford: Oxford Institute for Energy Studies, 1999), p. 33.

of gas resources will take the initiative, developing transportation links to get to the market ahead of their competitors, or will wait until the market 'becomes attractive'.

Privatization and liberalization of gas markets

During the 1990s gas markets worldwide entered a new era. Changing national and international agendas have produced a strong trend towards market liberalization. A desire on the part of governments to reduce fuel prices – principally to industrial customers – was the major driver of change. These changes were seen first in the United States and Canada, where 'deregulation' of traditional gas industries resulted in the removal of gas price regulation, open access to gas infrastructure, and the development of short-term trading with spot and futures markets. Pipeline companies were required to separate their merchant and transportation functions, to 'unbundle' their transportation services and offer these on a non-discriminatory basis. Traditional long-term contracts with take-or-pay clauses did not disappear but were significantly modified with prices tied to the futures market. In the UK and Australasia, the privatization of gas and electricity utilities has seen similar trends develop. There has been a sharp increase in gas demand, especially in the power sector. Privatization and liberalization have led to gas-to-gas competition, mutual 'invasion' of markets by gas and electricity utilities, and even multiple utility companies (electricity, gas, water and telecoms).

But there is significant diversity both within and among regions. Not all countries have followed – or wish to follow – the much-quoted examples of North America and the UK. Different levels of gas demand and stages of infrastructure development are the principal sources of diversity. In markets with established gas networks, the focus is on establishing conditions within which competitive commodity gas markets can flourish. In markets where the necessary infrastructure is lacking, the creation of competition may need to take second place to establishing large-scale networks so that the industry can create a critical mass of demand.

The move towards competitive and liberalized markets may not necessarily be possible for countries lacking the network infrastructure. It is worth remembering that the infrastructure in both North America and the UK was largely installed under monopoly regimes which provided security of throughputs and returns on infrastructure investment. In these mature markets, monopolies were abolished but only after much of the required network investment had been completed. It remains to be seen whether and how, in countries where the majority of network investment is still required, liberalized market arrangements can ensure rapid development of infrastructure. Alternatively, some transitional link will need to be made between the country's existing, or slowly liberalizing, market structure and the international investment markets for funds.

The industry perspective and commercial challenges

The simultaneous liberalization of the electricity and gas industries raises the possibility of downstream integration into electricity by gas, as is already occurring in North America and the UK. At the same time, the gas industry is restructuring itself to meet future challenges, which have significant potential impacts on the way in which future natural gas investment takes place. The wave of mega-mergers – BP/Amoco/Arco, Exxon/Mobil, Total/Fina/Elf – has both diversified the gas investment opportunities for companies and provided a significant capital base for large gas projects. For example, BP has become more of a gas company with the Amoco and Arco mergers, Exxon is combining its Asian gas resources with Mobil's Asian LNG experiences, and Total is becoming a player in LNG in Latin America and the Atlantic basin to supplement its strong Pacific basin position.

The coincidence of efforts to restructure the gas industry and advance investment in gas infrastructure will prove challenging, especially as the capital expenditures involved are enormous. However, policy efforts are being made to ensure this happens. The future organization of the industry will test the theories that mega-mergers facilitate capital investment and that downstream integration into electricity is a trend that is well under way and will accelerate in the future.

Environmental issues

New environmental regulations in the United States have helped to revive natural gas consumption. A similar trend is spreading worldwide at regional and local levels. Improving urban air quality is an important driver of Chinese policies aimed at increasing gas usage. The economic attractiveness of gas relative to coal increases when costs are attached to the emissions from burning these fuels. With gas, negligible SO_2 and significantly reduced levels of NO_x are produced, compared with fuel oil and coal. In the United States, generators now have to cap their SO_2 emissions or buy permits, and are finding as a result that new gas-fired plants are preferable to coal even with high-cost gas ($3/mm Btu). In combined cycle gas turbine (CCGT) plants, power generation reaches efficiencies of 50–55%, compared with a maximum of 40–45% for conventional plants.[36] Low ground-level pollution means that new gas-fired power stations are easier to site closer to electricity users. This facilitates the development of combined heat and power (CHP), or co-generation plants, where the steam is used for industrial processes or heating.

Table 5.13: Emission characteristics of natural gas

	G/kwh delivered electricity		
	Gas combined cycle	Fuel-oil power station	Coal power station
Efficiency %	58	40	40
CO_2	313	673	813
NO_x	1.04	1.73	2.7–7.2
SO_2	0	1.7–5.0	2.3–7.2

Gas produces less than half the CO_2 emissions per unit of electricity produced from coal. The switch to natural gas envisaged in the 'conventional wisdom' could be driven further if strong policies are adopted as a result of the UN Framework Convention on Climate Change (UN FCCC). The Kyoto Protocol to the UN FCCC, which was agreed by the Third Conference of the Parties (COP3) in December 1997, set a

[36] 'Clean coal' technologies will narrow the gap. See IEA, 'The Clean Use of Coal', in *Energy Technologies for the 21st Century* (Paris: IEA/OECD, 1997), ch. 2.

ceiling on the amount of emissions of GHGs from each industrialized country for the period 2008–12. In the event of these commitments becoming legally binding, the outlook for coal in industrialized countries could deteriorate if the targets were implemented by policies which targeted emission reductions according to the carbon/energy ratio of the fuel. However, at the time of writing in June 2000, it is not clear when ratification is likely. Current trends for emissions in most countries are far from the targets. (This question is discussed more fully in Chapter 8.)

Market horizons for new technologies

Fuel cells (for transport and stationary use), and micro power generation are attracting attention as technologies capable of having a substantial impact on the expansion of gas markets during the period 2010–20. Gas-to-liquids technologies may also enable owners of remote gas resources to bring gas into the liquid fuels market and avoid costly gas transportation. Many of these technologies are not strictly 'new' – they have all been under consideration for many decades. But technical and commercial breakthroughs can be expected within the next 10–20 years which would have a very powerful impact on gas demand and supply. Micro power generation plant which could be installed in buildings to provide combined heat and power to the building and to a grid, or fuel cells which could do the same, would unite the gas and electricity markets in an entirely new way. Improvements in the storage and distribution of CNG and liquefied petroleum gas (LPG) for use in the transport sector could also contribute to the diversification of gas markets into high-value uses where alternatives, such as gasoline or diesel, now carry heavy environmental penalties for their SO_2 and NO_x pollution.

Conclusions

The paradox of 'increasing demand and increasing prices'

In conclusion we return to what has been referred to in Chapter 2 as the 'paradox of increasing demand despite increasing prices'. There is a relative abundance of material in the general energy and natural gas literature on reserves and resources, production potential and potential infrastructure projects. But there is a tendency for these data to be used to generate demand projections without sufficient reference to prices: whether producer (wellhead or landed/beach) prices, exporter (border) and end-user (power generation, industrial and residential/commercial). The scenario literature does contain price projections, but only for three main OECD gas market regions: US wellhead (or import) prices, European import prices and Japan (or Asian) LNG import prices.

Table 5.14: Comparison of gas-price scenarios

Gas prices (1999$/mm Btu)	2000[a]	2010			2020		
		IEA	POLES	EIA	IEA	POLES	EIA
US wellhead	2.31	2.10[b]	3.80	2.72	4.35	3.91	2.92
European import	2.54	2.99	3.24	N/A	4.29	4.23	N/A
Japan LNG	4.26	4.03	4.67	N/A	6.04	4.99	N/A

[a] Actual January 2000 prices: US Futures (NYMEX), German import average (first quarter 2000), Japan LNG import weighted average.
[b] 1998–2005.

Source: Data in Tables 5.5–5.6, converted to 1999$ and mm Btu; see note to Table 5.5 above.

Table 5.14 shows a comparison of the three gas price scenarios. With the exception of the IEA 2010 Japan/Asian market price, they all suggest a significant increase in prices compared with those of early January 2000.[37] What is interesting about the price scenarios is that they go hand in hand with demand scenarios which also show significant increases over the same period (see Table 5.1). The only exceptions to this are the IEA projection which shows North American gas demand falling during the period 2010–20, owing to higher prices; and the EIA price scenario which shows North American residential prices falling during

[37] The choice of January 2000 is somewhat arbitrary but is intended to represent a midpoint in the price cycle. At the time of writing (June 2000), strongly rising oil prices over the previous six months accompanied by a tightening US supply/demand balance have pushed prices up considerably to $3/mm Btu in Europe, $4.40/mm Btu in Japan (February 2000) and $4.28/mm Btu in the United States. This compares with considerably lower prices in mid-1999: Europe $1.63/mm Btu; Japan $3.06/mm Btu; United States $2.22/mm Btu.

the period 1998–2000 (see Table 5.7), presumably under the pressure of market liberalization.

It *may* be possible to generalize the regional price data to the whole of 'North America' (United States, Canada and perhaps Mexico), 'Europe' (western and eastern); and existing Asian LNG importers. However, these price scenarios are of limited use for large parts of the world, such as economies in transition (FSU), Latin America, Africa, South and East Asia. Since it is in these regions that more than 60% of the global gas demand increase is projected to take place over the next two decades, this is a significant gap in the analytical framework. In many – perhaps most – of these regions, prices would be expected to increase substantially over the next two decades, due to the removal of subsidies and the enforcement of payment for gas, and this must place question marks over scenarios which see large demand increases over the same period.

Quite understandably, because of the very great uncertainties, and data problems, which such an exercise would entail, there has been little attempt to forecast producer or border prices for markets outside the OECD, or end-user prices outside North America. Such problems are compounded by the wave of liberalization sweeping through natural gas (and electricity) markets, which adds another level of uncertainty. The traditional commercial structure of long-term take-or-pay contracts containing prices with oil-based indexation, applied to infrastructure over which owners have monopoly control, can no longer be assumed. The forecasting problems associated with liberalization are sufficiently complicated in OECD markets where pipeline infrastructure, price mechanisms denominated in convertible currencies, and (some degree of) regulation, are established. But in many countries and regions outside the OECD, any or all of these elements may be absent.

In countries such as Russia (and other transition economies), China and India, liberalization will almost certainly require a drastic increase in prices – as subsidies are removed and payments are rigorously enforced – rather than the decline in prices which has been the experience in North America and the UK. Price increases may help to support the feasibility of higher-cost projects but they are likely to curtail the demand (particularly) of end-users outside the power generation sector.

This may be especially relevant where new low-pressure reticulation systems need to be built, as opposed to converting existing town gas grids. Where liberalization renders it impossible for long-term take-or-pay contracts to underpin investments, this will also complicate projects requiring multi-billion-dollar greenfield supply infrastructure in developing country markets.[38] Such infrastructure will be a requirement for most Asian and Latin American countries and will determine both the extent and the timing of demand increases in these countries.

OECD countries seem likely to continue to move towards liberalized markets, with prices influenced by gas-to-gas competition as well as by the prices of competing fuels such as oil – provided the competitive market exists. In the United States short-term competition is provided in the very large number of industrial plants with the capacity to switch fuels in existing combustion equipment. In Europe (and the United States) it is provided by the capacity of power stations linked in a grid (and with competitive electricity pricing) to switch the electricity load between gas-fired turbine stations and coal or oil stations. Without such conditions, the role of oil prices may diminish. When switching capacity is exhausted in the short term, gas prices may be determined by gas-to-gas competition in the supply–demand balance for gas alone. In the medium term crude oil and oil product prices will provide a 'ceiling' above which gas prices cannot remain for more than a short period (and, reciprocally, gas prices will provide a ceiling for oil). Depending on environmental standards and costs, coal prices may provide a 'floor' for both petroleum fuels in the power market (although this is less certain). But outside OECD countries, and especially where natural gas is being introduced into countries for the first time on a large scale by means of a single source flowing through a single pipeline, prices of competing fuels – and especially oil – may remain closely linked with gas prices.[39]

---

[38] The case for a continuation of traditional long-term take-or-pay contracts to support new infrastructure in OECD countries is substantially less strong.

[39] Gas-to-gas competition may well prevail in non-OECD regions such as economies in transition which are already highly gasified. In addition, oil prices may not be the determining factor in highly coal-based economies such as China and India.

These factors suggest reason for greater caution in devising scenarios which show large increases in both gas demand and gas wellhead/import prices over the next two decades, unless more attention is paid to where these suggested price increases will fall in the value chain. The fact of a huge established resource base suggesting availability of large volumes of gas for a wide variety of countries and regions, and the considerable environmental advantages of gas over other fossil fuels, are clear. But a major question that arises from this chapter is whether the price increases which will be needed to commercialize the gas resources will accrue to producers or to those further down the value chain.

The proposition of many owners of established resources is that they may have 'insufficient' incentive to commercialize them at prices which provide a relatively low netback at the wellhead. But for a large proportion of discovered reserves and resources in countries and regions with relatively small populations – for example, in the Middle East and the Siberian Far East – there is no conceivable alternative to export markets in Asia. If the owners of those resources do not have the incentive, the organizational capacity and the financial backing to develop them for the next step in the expansion of export markets, then it should be assumed that their competitors (both gas and non-gas) will take advantage of such inaction. For owners of undeveloped gas reserves, the present value of their assets will depreciate for the next two decades, as it has for the past two.

From this analysis, scenario assumptions that wellhead and border gas prices will inevitably increase in real terms over the next 20 years require serious re-examination. These assumptions give producers and resource holders potentially unrealistic expectations that they will be progressively 'in the driving seat' of natural gas development over the next two decades. A more likely conclusion from this analysis is that prices paid to gas producers will not increase over this period and may decline. The only region where this does not hold true is North America, where the level of reserves is sufficiently low to create competition between imported oil and imported gas. The latter would enjoy some environmental premium which might lead to an increase in US wellhead prices.

As far as prices paid by end-consumers are concerned, a distinction is required between OECD and non-OECD markets – or perhaps a better distinction would be between gas markets with significant gas penetration and network infrastructure, and those with less well-developed (or undeveloped) networks. For developed OECD markets, liberalization and gas-to-gas competition will reduce prices to all groups of end-users – perhaps significantly – and will maintain downward pressure on prices.[40] For a large number of less developed (mainly non-OECD) gas markets, end-user prices will need to increase significantly the customer end of the value chain in order both to eliminate subsidies and to finance the building of new transmission and distribution infrastructure. However, such price increases are unlikely to increase the netback to producers/exporters to those markets, as most of the additional revenue will be swallowed by transmission and distribution infrastructure requirements.

To the extent that producers take the view that unless they receive higher prices they will be unwilling to develop resources that they have already spent money identifying. This will curtail the potential for gas demand which exists in almost every region of the world. It will also present opportunities for other fuels – not excluding oil – to retain or increase market share to a degree not envisaged in the 'conventional vision'.

What seems certain is that the emphasis of the next two decades in the gas industry will be on financing and building gas infrastructure to develop gas markets with end-user prices that will ensure expansion of demand. In order to achieve this, there must be a clear expectation that:

- customers – whether power plants or other end-users – are able and willing to pay these prices; and
- gas prices will be significantly below those of competing fuels (taking account of environmental premiums or penalties) on a sustained basis.

[40] The exception to price reductions may be in markets such as the United States, Canada and the UK where competition is already established – particularly in non-residential sectors – and some increase in prices is at least possible.

These expectations clearly depend mainly on the prospect of continuing significant reductions in the cost of gas transport and distribution before there is any prospect of improving, or even maintaining, wellhead prices.

The expansion of demand foreseen in most scenarios therefore depends in many cases on *both* significant reductions in transportation and distribution costs *and* – in some cases – sufficiently high wellhead prices to expand production. In some cases, the size of the undeveloped reservoirs provides very low production costs: there the question is how long the resource owners are prepared to wait to develop their reserves. In general, it is thus cost and the owner's expectations that will be the main battlegrounds for gas over the next two decades. Discovering new resources is not important, unless they are near to the potential markets, such as China and India.

Outlook

Over the next two decades, it seems probable that:

- gas use will continue to grow in mature markets (but not in all sectors) and very significantly in new markets – to some extent at the expense of oil as well as coal. The possible exception may be North America after 2010;
- like coal prices, gas prices will be one of the girders on which the ceiling to oil prices will rest in future;
- an increasing part of gas consumption in most regions will be supplied by imports; and
- there will be increasing export opportunities for governments and companies with large gas resources, and with the finance, technology and skills to bring them to markets, but they will remain in competition both with one another and with the promoters of other fuels.

Along with these advantages will come a number of challenges:

- Resources will need to be moved to markets from locations increasingly remote from those markets.

- In much of Asia, the development of large-scale markets will depend on whether the infrastructure is created to move these resources to centres of demand.
- The privatization of state-owned companies and liberalization of gas network infrastructure cannot be automatically assumed to be favourable to such developments. Financing projects in markets with little established gas infrastructure may be further complicated if liberalized access to networks is envisaged from the outset of deliveries. The liberalization requirements of multilateral lending agencies may be a further factor complicating multi-billion-dollar financing.
- Gas market development will be strongly allied to the development of power generation. Privatization and liberalization developments in the electricity industry will therefore be of great significance for gas markets.
- While environmental issues will be considered a generally positive feature of new gas developments – especially in terms of alleviating local emission problems – there will also be negative aspects of land disturbance and issues of governance.
- Rapid and successful development of supply- and demand-side technologies, e.g. gas-to-liquids and fuel cells, may change dramatically the calculations of available gas supply and demand. Whether this will happen prior to 2020 is uncertain.

Meanwhile, boundaries between the private and public sector, and between the oil, gas and electricity industries, are shifting and becoming more permeable. The introduction of market competition changes the focus of regulation but does not eliminate it. A judicious mix of both competition and cooperation is required if a sustainable future is to be built. The trend to privatization and the promotion of open competitive gas markets is offering openings at the same time for international oil companies to find new source of growth and profit, and for national gas and electricity companies to internationalize and diversify their risks. This new 'jungle' will hold threats for some existing players – particularly in countries with weak companies or restrictive governments where resistance to change carries heavy economic costs.

The issue of supply security remains high on the policy agenda, particularly for governments with high and growing import dependence. Thus there is clearly a role for industry and a role for the state.

Chapter 6

Oil prices: the elastic band

This chapter proposes a way of thinking about oil prices in the 'new economy' of oil. The key idea is that, although they are important, oil prices are no longer central to world energy prices. We are now in the 'long run' after the shocks of the 1970s. We need to think again.

The first part of the chapter highlights some facts from the epic history of oil prices. A credible theory for building scenarios for future oil prices should try to explain the past. Over the very long run – the past century or more – it is difficult to discern a steady trend: the record is rather of 'episodes'. These correspond to changes in expectations about future oil reserves and markets, and changes in the role of governments, either affecting ownership of the resources or intervening in the structure of the market. Econometrics supports the idea of episodes because it appears that oil prices revert to a trend, but that the trend changes.

The chapter then goes on to sketch a relationship between these 'episodes' and two academic approaches to explaining long-term oil price movements:

- The 'exhaustible resources' idea that oil producers discount the value of oil kept 'in the ground' for future production, and therefore are prepared to accept lower prices today than they expect in the future. The idea is important because different exporting producers have very different reserves of oil in relation to their current production capacity: these differences drive competition to expand production capacity and then production.
- Models about how the international market price can be influenced by a core cartel of major low-cost exporters (or a single dominant 'market leader') facing competition from minor producers with lower reserves but higher costs. Because these major exporters are governments, there is a 'survival' level of prices at which they agree

to restrain the competition into which their otherwise different interests drive them: this sets a lower band for oil prices.

This is followed by a discussion of the setting of the next 'episode' – the 'new economy' of oil. There are two key ideas:

- Competition from other fuels – especially natural gas – has been intensified by changes in the regulation of electricity and gas markets, and by emerging competition between transport technologies: the 'long term' in which these set an upper band for oil prices is rapidly becoming shorter.
- The reduction of the massive structural surpluses of oil production capacity created by the oil price shocks of 1973 and 1979 will simplify the task of restraining competition when prices are at the 'survival' level.

The conclusion is reached that 'market leadership' between upper and lower limits is unlikely to control the competitive forces within or outside the oil market, so that the oil price in the short-to-medium and medium term will be indeterminate – the subject of chance and transient factors. The idea of intermittent competition among the major exporters with market power, and continuous competition from alternatives to oil, is central to this section.

The oil price epic

The history of constant-dollar oil prices is difficult to interpret by means of a simple story. This is not surprising. Oil demand may be a relatively predictable function of economic activity, the price and availability of alternative fuels, and the price of oil itself. The supply of oil is subject to other factors which have, in the past, had a more dramatic and unpredictable character. One storyline is the changing perception of the future balance between oil demand and the resources available for future supply. Another is the changing role of governments in the oil industry. Linking the two have been *coups de théâtre* in the form of the oil shocks, sparked by some combination of events such as war, revolution, unusual weather, economic cycles and changes of government. Many of these have changed the structure of the oil market for a period, as well as the short-run price. Some of the same factors have affected oil's principal competitors for supplying fuel to power stations, industries, shops and homes in the main consuming countries. For the competitors, what hurts oil may help them.

Changes in demand

Demand for energy has generally been linked with economic growth, as described in Chapter 2. Technology changes the link.

The demand for oil as a *transport* fuel has been affected by particular technical developments. Since the 1920s the availability, price, reliability and comfort of motor vehicles has dramatically expanded the demand for liquid transport fuels – a trend modified somewhat in the United States by the CAFE standards introduced in 1980.[1] Chapter 4 sets out reasons why this trend may be about to change.

The growth of *electricity* demand has been driven by the falling cost of household electrical machinery and appliances, as well as of much industrial and commercial equipment; better control of the use of this equipment by electronics is now probably modifying this growth. There are many substitutes for oil in the power market. The development of *nuclear* power technology enabled the expansion of nuclear power in the 1970s and 1980s. Improvements in *combined cycle gas turbine* technology have made gas a fuel of choice for new power generation in Europe in the 1990s. However, these improvements have yet to work their way through the transport and power generation systems of many developing countries (see Chapter 5).

The application of new mining technology, and in particular the better management of mining and transportation processes through the use of electronic controls, has enabled major *coal* producers (such as the United States) to continue to reduce costs. With falling prices and increasing production, the coal share of the electricity market in industrial

[1] The US government required automobile manufacturers to reduce the fuel consumption average of their fleet of new car sales. See Chapter 4.

countries as a group, has stabilized and is expected to be maintained around current levels, although it has declined in Europe.[2]

The long-term effect of such changes, through a variety of other factors affecting the industries concerned, has immediate impact on expectations for future oil demand. Even oil companies have learned that there are substitutes for oil. Between 1973 and 1985, a period of high prices, oil lost 10% of the world energy market that it had gained in the preceding period of falling real prices.

Changes in perceptions about future resources

More (or less) oil in the future means that the owners of reserves today will expect lower (or higher) prices in the future and may prefer to produce more (or less) at today's prices. Changes in perceptions of reserves may therefore change today's prices. The historical record of price changes (see Figure 6.2) roughly mirrors alternating perceptions of abundance and scarcity as new areas were explored.

In the first half of the twentieth century, perceptions of abundant oil resources were supported by major discoveries: first in Iran and Mexico at the beginning of the century, then in Venezuela and Iraq in 1928, followed by Saudi Arabia and Kuwait in 1938. The 1950s and 1960s saw discoveries and developments in Nigeria, Libya and Abu Dhabi. In each case the discovery (or, in the 1950s and 1960s, the continuous trend of discoveries) was followed by a downward movement in oil prices – before the discovery had a major impact on production.

In 1959, in the face of rising supply, the companies, led by Exxon (then Standard Oil of New Jersey), unilaterally dropped the price that they posted (for tax reference purposes) for exports from the main exporting countries: an action which prompted the formation of OPEC by the key governments of exporting countries.

By the end of the 1960s the perception had turned to one of scarcity. Prices had been falling in real terms. Oil was displacing coal in many markets. Oil consumption was rising faster than reserves. There was competition among exporters to increase liftings. The rate of new

[2] EIA, *International Energy Outlook 2000*, table 21, p. 115.

discoveries was falling. The costs of developing the new discoveries in offshore areas of the North Sea, Brazil, Mexico and Alaska were expected to be high. By the time the companies and exporting governments began to renegotiate posted prices in 1970 there was an expectation that the growth in supply could not be sustained. This impression was strengthened by the decline in US production that began in 1970, because of a lack of reserves. For the private sector producers, the situation was made worse because their interests in many oil-exporting countries were threatened by the end of their concessions, by nationalization, or by participation. 'Their' oil reserves collapsed.

Expectations of future 'scarcity' subsided during the late 1980s, at least among those in the oil business. This was not because of major new discoveries of resources – although substantial new discoveries were still being made offshore. The rate of new discovery still fell far short of the rate of depletion of existing reserves. But reserves continued to grow (see Chapter 3, Figure 3.3). New computing and communication technology was imported from the rest of the 'new economy' in to the key development activities of reservoir modelling and the control of drilling. Lower oil prices from 1986 made developers and engineering contractors concentrate on reducing costs. This new focus led to large and apparently continuing reductions in the cost of production in the private sector and therefore to the upward revision of the reserves recoverable from known resources. The idea of 'reserves growth' implied that similar improvements could be expected from the known resources of the state sector as a result of the increasing use of private sector companies as contractors.

Governments changing the rules

There were epic changes in the ownership of oil during the twentieth century. Up to 1971, 80% or more of oil exploration, development and supply of crude oil was in the hands of private sector companies. By 1980 their share had fallen to around 30%, recovering only slowly during the 1980s to nearly 40%. During the 1990s the distinction between private and state ownership became increasing blurred, but it is

Figure 6.1: Shares of world oil production

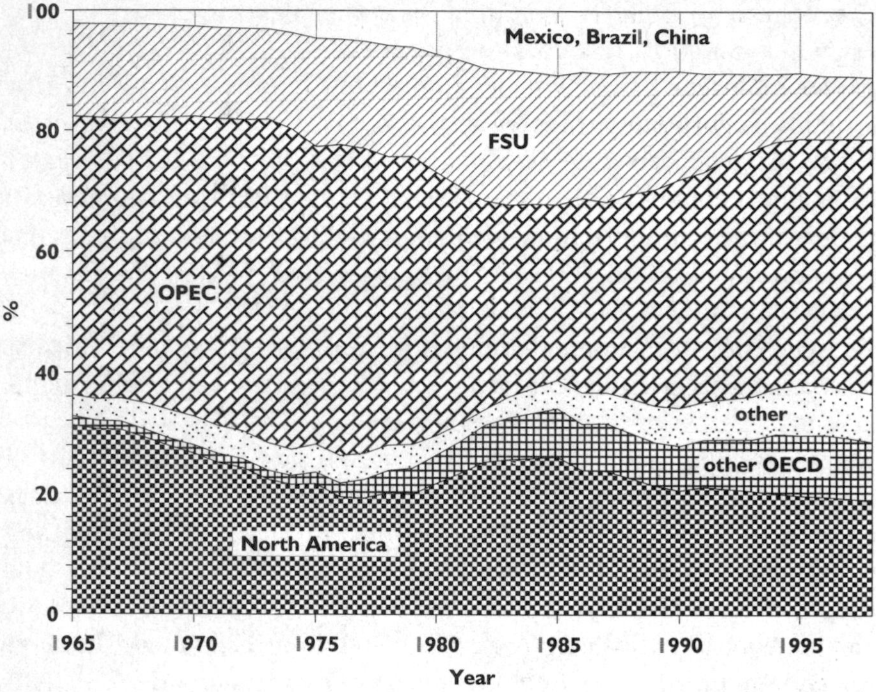

Source: BP Amoco *Statistical Review 1999.*

possible to say that in 2000 the private sector is involved in some way in about one-half of world oil production.[3]

These changes are illustrated in Figure 6.1. During the period covered by the figure, US and almost all Canadian production was in the private sector, and Brazilian, Chinese and Mexican production was entirely in the state sector. Elsewhere, ownership was mixed. In most OPEC countries, concessions for private sector exploration and production

[3] Exact definitions are not possible. Some OPEC members, such as Indonesia, Nigeria and Libya, never completely nationalized the oil industry in their countries, but control its development through 'participation' – state companies taking a majority share in joint operating companies. For a description of the *aperatura* in Venezuela, see Bernard Mommer, *The New Governance of Venezuelan Oil*, WPM23 (Oxford: Oxford Institute of Energy Studies, 1998).

were eliminated, and in some OECD countries with growing production, such as Norway, there was a substantial state participation.

Various factors determined these shifts:

- The transfer, at the insistence of the governments concerned, of total or majority ownership of the national oil industry to government companies in most OPEC countries during the 1970s.[4] There was a political context to this. The US and UK governments did not contest these nationalizations as they had done in the case of Iranian nationalization in 1951–53, for by now their influence in the area had lessened: Egypt had defied the British and French over Suez, the second American intervention in the Lebanon had failed, British troops and ships had withdrawn from the Gulf for budgetary reasons.
- The decline (because of depleting reserves) of the importance of North America in oil production since the Second World War, from producing two-thirds of world production in 1950 to under 20% today.
- The rise and fall of production in the Soviet Union during the 1980s, followed by the break-up of the state monopoly and privatization in the mid-1990s.
- The growing diversification of supply, partly in countries where state monopolies apply (Mexico, Brazil, China) and partly in countries where the private sector operates under 'production-sharing' contracts. Many OPEC countries have in the 1990s reintroduced private sector companies as contractors or as partners in production-sharing contracts for new developments or for 'rehabilitation' projects.[5] The reserves involved are small in relation to the countries, but large in relation to the companies concerned.

---

[4] The concessions under which private sector companies operated in these countries did not in general amount to ownership: development plans and even production targets were in many cases subject to negotiation with the government. Prices, for the purpose of local royalties and taxation, were set until 1973 by negotiation. Nevertheless, the change was profound.

[5] Contracting arrangements of various sorts were reintroduced to Venezuela and Algeria during the late 1990s.

Before the changes of ownership, investment and production policy in most exporting countries had been set – subject to some negotiation with individual 'host' governments – by the major private international companies, most of whom had parallel interests in several exporting countries. From 1928 to the Second World War, competition among the companies for upstream resources was limited geographically by the 'Red Line' agreements reached by some of the major companies. In 1947 the 'Red Line' agreement was abrogated; there was a reorganization of concessions in Saudi Arabia, Kuwait and the areas of operation of the foreign-owned Iraq Petroleum Company. The result was a web of cross-interests. In each major exporting country there was generally a main concession of exploration and production rights by the government to a consortium of companies:[6] these companies played a role in matching total supplies to demand and in moderating competition among exporting countries.[7]

By the early 1980s investment and production decisions in most major exporting countries were taken by the governments and carried out by state companies. Production quotas were from time to time set on a short-term basis through OPEC, although Saudi Arabia, the principal producer, has refused as a matter of principle to consent to a quota being agreed for its production. More recent developments are described below.

Changing regulations to protect producers

In the United States and Canada, periods of excess supply during the 1930s led to production rationing in the principal oil producing states: under the US Oil Code of 1933 the Secretary for the Interior could, and

[6] Iran was the main exception. The main concession was to the Anglo-Iranian Oil Company (later BP) until 1951, when its interests were nationalized by the Mossadeq government. The concession was restored in 1953 after a coup engineered by the CIA, but to a consortium in which the BP interest was limited to 40%, with US companies holding the balance. However, Anglo-Iranian and BP entered into long-term sales agreements to supply Iranian and Kuwaiti crude to Shell and other major oil companies, whose downstream demand was thus incorporated into the planning of supply in those countries.

[7] For an analysis, see J. E. Hartshorn, *Oil Trade: Politics and Prospects* (Cambridge: Cambridge University Press, 1993), pp. 191–2.

did, specify production quotas, and state utility commissions did so on a state level. The growth of cheap imports in the 1950s led to federal import controls to protect domestic producers.

The price explosion of 1973 led to price controls on domestic production, the allocation of price-controlled supplies to refiners, and excess profit taxes on 'old' oil (as well as gas). There were similar moves in Canada. In the UK and Norway special tax regimes were imposed on domestic production. Within a decade these upstream interventions in North America had disappeared. In the macroeconomic and political environment of the 1980s in the United States and the UK, the purpose of policy was to reduce government intervention in the marketplace. Price controls in Europe did not long survive the disruptions of 1979–80 and by 1998, even Japan was dismantling its regulation of petroleum imports, refining and distribution. There were parallel movements affecting electricity and gas in all these main consuming regions and in some developing countries as well, under the influence of the IMF and the World Bank.

Changing regulations to protect consumers

Although governments have withdrawn from the attempt to manage markets and supply, they did, over the century, intervene more and more for other public purposes: markets have been regulated, not to stabilize prices or protect investment, but to increase competition. A US judge dissolved the Standard Oil Trust in 1911. In the 1980s the United States began to change the regulation of privately owned monopolies in airlines, railways, power and gas distribution. Where one element in a value chain was a 'natural monopoly' – as in pipeline transit, power transmission, telephone lines and railways – regulators required it to be segregated to some degree from the other operations so that its profits could be isolated and access be visibly made available to competing users. The other elements in the chain were open to competition. This 'unbundling' (to use the language of utility regulation) of integrated functions is now the fashion in industrial countries. The de-integration of the oil industry is remarkable because it was one of the few examples of an industry that had been integrated across frontiers. As a result of OPEC's actions in the 1970s it 'de-integrated' abruptly, 'unbundling' to a degree and with a speed which would probably be impossible to achieve by slowly crafted international agreements to promote competition.

Changing regulations to protect the environment

OECD countries have also seen a trend towards more government intervention to protect human health and the natural environment. In most OECD countries environmental protection agencies have been delegated regulatory powers. These agencies prescribe product qualities and manufacturing and distribution processes in great detail with the objective of avoiding or reducing polluting events such as emissions to the air, spills on the ground and leakage into water systems, among other contingencies. Investments of many kinds require examination of environmental impact before government or agency approval.

The United States, in Title IV of the 1990 Clean Air Act, launched and created a new commodity and a market for it: permits to emit sulphur dioxide. This example may set the trend for allocated tradable permits or allocations of rights to emit CO_2 as climate change policies are developed in response to the UN Framework Convention on Climate Change of 1993 and the Kyoto Protocol commitments of 1998.

The 'polluting event' is often the result of the combustion of fuel in the vehicles, buildings or equipment owned by final consumers. Avoiding pollution, however, requires clean fuels as well as efficient burning. The regulation of fuel quality has thus become a major objective in legislation such as the US Clean Air Acts of 1980 and 1990 and the EU directives on combustion plants. The EU, in its auto-oil programme and the resulting directives for the automobile and refining industries, has pioneered the idea of seeking to develop a cross-industry regulation of fuel and equipment to achieve specified air quality objectives by the least-cost combination of fuel and equipment.

Shocks and starts

In explaining the history of oil prices it is a major challenge to distinguish the normal – itself the result of a combination of conflicting trends – from the abnormal and transient. If the abnormality of the peaks in 1979–81 had been recognized, forecasters might have been less likely to extrapolate from those events as a basis for future projection. The same goes for the troughs of 1986 and 1998.

The 'oil shocks' of 1973 and 1979 combined transient and temporary disruptions of supply with unstable or changing political as well as oil industry circumstances. The 'oil shock' of 1973 was symbolized by the attempts by Arab oil producers over a few months to embargo companies alleged to be supporting Israel in the Yom Kippur War. This coincided with strong cyclical growth in demand and special weather conditions in importing countries, as well as critical turning points in the perceptions of reserves and the relations between the international oil companies and the governments of exporting countries. Likewise, the Iranian Revolution and the Iran–Iraq War of 1979 and 1980 seemed to be a climactic demonstration to importers of the economic dangers of depending on the international oil trade because prices rose so far and so fast. In fact, supplies were not disrupted for long. The disruption did not exceed the 7% prescribed as the automatic trigger for IEA action, emergency sharing schemes and the drawing down of strategic stocks. Many of the importers' difficulties were attributable to their governments' failure to coordinate intervention at a time when international oil trade had not yet been commoditized.[8] As in 1973, the political context was unfavourable and uncertain: the intentions of the Iranian revolutionaries were unknown. The failure of the United States to solve the crisis of the American hostages imprisoned in its embassy in Tehran seemed to confirm the message of the 1973 events: the exporting governments were in charge now.

The disruption occasioned by the Gulf War of 1990–1 did not correspond with any fundamental change in resource expectations,

---

[8] For an account of the mismatch between government interventions and the realities of supply, see Daniel Badger, 'The Anatomy of a Minor Disruption', in Alvin and Weiner Ribert, *Oil Shock: Policy Response and Implementation* (Cambridge, Mass.: Dallinger, 1984).

ownership, market structure or technology: in fact, the war was fought essentially to prevent further changes in ownership. The effect on prices was transitory. Saudi Arabia increased production to make up for lost supply. The IEA coordinated OECD importer government actions, and at a critical moment strategic stocks were released to reassure the market.[9] The political context – and outcome – were entirely different from 1973 or 1979. In 1986–7 the United States had successfully intervened, with naval help from the UK and France, to protect shipping in the Gulf from Iranian attacks and to exclude any possible Soviet intervention. By 1990, the United States was the dominant power in the Gulf. The collapsing Soviet Union used skilful diplomacy to keep a stake in a game in which it could not possibly intervene. The outcome, disastrous for the Iraqi people, was the establishment of permanent US bases in the Gulf, backed in the case of Kuwait and the UAE by formal security agreements.

Market structure changes

The stories of changing expectations about resources, changing government intervention and dramatic shocks provide a working explanation of the 'episodic' history of oil supply. Three phases stand out:

- the *integrated market* which prevailed until 1971;
- a *transitional period* between the first and second oil shocks; and
- the *commodity market* which has prevailed since 1980 and whose reach is spreading into the emerging market economies and developing countries as they deregulate their markets.

Oil prices were determined differently in each phase.

During the 'integrated phase' up to 1971 most international trade flowed either within a few major private sector companies or among them:[10] prices had been set by negotiation of 'tax reference' or posted

[9] For an account, see Peter Huggins, 'The IEA's Response to the Oil Crisis of 1990–1', paper presented to the Commodities Futures and Options Association, Burgenstock, 1991.

[10] This is a generalization: the allocation of concessions in Libya and Nigeria to a number of smaller (mainly American) non-integrated companies had already put 'third party' oil on

prices between these companies and the major exporting governments.[11] This system began to crumble during the Libyan government's negotiation with successive companies during 1970–1, and the subsequent 'Tehran' agreements between Iran and other OPEC producers and a group of concessionary companies.

The 'transitional phase' began abruptly in 1973: negotiations broke down in the face of OPEC demands for a doubling of posted prices. At the same time a group of Arab oil governments imposed an embargo on exports to countries they associated with US military support for Israel when it was attacked by Egypt at the beginning of the Yom Kippur War. In October 1973 OPEC governments announced that prices would in future be set unilaterally.

During the 1970s the transitional arrangements of 'participation' continued to keep a large, though diminishing, fraction of trade within long-term contracts between the 'participated' state-dominated companies and the former concessionaires. Long-term pricing still seemed possible; special terms for former concessionaires continued to exist.[12] This system was incapable of handling disruptions. There was no liquid spot market to which customers affected by the disruption of Iranian supplies could turn. There was no futures market in which traders could indicate their views of future balance of supply and demand and in which speculators could add liquidity. Nor were there the rigid supply contracts between the major companies and their customers which, in the 1973 oil shock, had automatically created a 'pro-rationing' of supplies. Prices for spot cargoes reached unprecedented levels.[13]

---

[10] (cont)

... to the international market. Some of this had originally been destined for the United States, but US import controls imposed in 1951 limited this opportunity. The independent producers exerted pressure on prices outside Europe through sales to minor European refining companies, particularly in Italy.

[11] This has not always been the case. The unilateral reduction of the posted (tax reference) price in 1959 by Standard Oil of New Jersey (later Exxon) had been a major reason for the creation of OPEC in 1960: see Daniel Yergin, *The Prize* (New York: Simon & Schuster, 1991), ch. 26, pp. 519–23.

[12] For a detailed history, see Hartshorn, *Oil Trade*.

[13] For a detailed analysis, see Badger, 'The Anatomy of a Minor Disruption'.

Figure 6.2: Oil prices since 1900

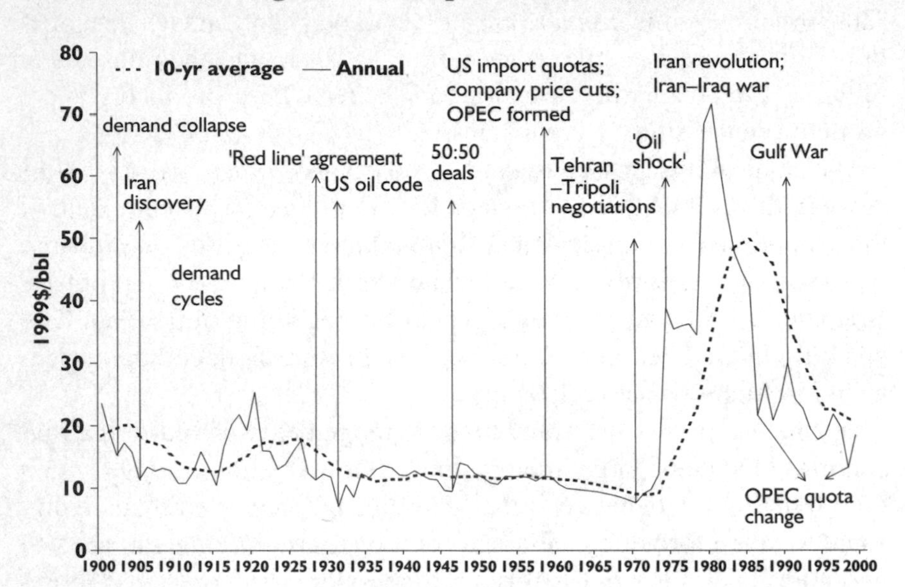

Source: Price data from BP Amoco *Statistical Review 1999* and IEA *Oil Market Report*, April 2000.

The 1979 oil shock brought the transition to an end. In 1979 OPEC governments abandoned the attempt to coordinate prices. Each country was now 'free' to set its own price. The 'commodity market' had arrived, by OPEC fiat. The long-term contracts collapsed; the New York Mercantile Exchange and the International Petroleum Exchange in London were established and developed crude oil contracts in the early 1980s. Various journals established systems for reporting spot prices of crude and products.[14] Although OPEC introduced production quotas in 1983 in an attempt to stabilize prices, competition among exporters led to the collapse of this experiment in 1986.[15]

To illustrate how these themes played through the history of oil prices, Figure 6.2 shows the annual average price compared to the ten-year

[14] See Paul Horsnell and Robert Mabro, *Oil Markets and Prices* (Oxford: Oxford University Press, 1993), ch. 10, pp. 155–66.

[15] See Hartshorn, *Oil Trade*, ch. 9; Dermot Gately, 'Strategies for OPEC's Pricing and Output Decisions', *Energy Journal*, vol. 16, no. 3 (1995), pp. 17–19.

moving average. The difference between the two is a simple indicator of 'abnormal' prices. Events marking 'episodes' are shown.

Episodes look like trends

There are thus a variety of practical explanations which suggest a story for oil prices which involves reversion – over a period which could be about a decade[16] – to a mean which may change from time to time. This is exactly the model which has been developed by Professor Pindyck.[17] In his paper he contrasts statistical results for fitting long-run oil prices (back to 1870) to three possible trends:

- a quadratic time trend: these estimates are very sensitive to the time period chosen;
- a fixed trend to which prices revert: though this fits 'better' than a time series trend, the extrapolations are still sensitive to the data period included – especially to the treatment of the high prices of 1973–84;
- a trend which fluctuates stochastically: this form of estimation would capture the 'epic' characteristics described above; the resulting 'predictions' – estimated from data up to 1981 – fit the actual prices from 1981 to 1996 better than predictions from a model which shows reversions to a fixed trend line.

There is no statistical or intuitive support for the idea of a long-run upward or downward trend line in the price of oil. Statistical models which show reversion to a varying mean appear to offer the best fit, and to be most in accord with the realities of the historic oil situation. Intuition suggests the idea of limits.

---

[16] See Robert S. Pindyck, *The Long-Run Evolution of Energy Prices*, MIT Center for Energy and Environmental Policy Research, Working Paper 99-001, January 1999, p. 7 and fig. 4. The paper is also published in *The Energy Journal*, vol. 20, no. 2 (1999).

[17] Pindyck, *The Long-Run Evolution of Energy Prices*.

Limits

The idea that there are limits or boundaries to the oil price is related to the idea that there are trends that vary. A trend that is the product of a particular episode may eventually generate a reaction. A rising price may cross the threshold at which a substitute like nuclear power becomes clearly the most economic way to generate electricity. A falling price may blunt the forces generating competition among producers at the point where all producers will be better off restricting competition than continuing it. In both cases the reaction is not continuous with continuously changing prices: it occurs when some new factor is triggered.

Changes in oil's share of the energy market reflect changes in the price of oil: a complete analysis would take account of changes in the price of other fuels and the relative efficiencies of combustion.[18] But it is easy to form a rough picture. Figure 6.3, like Figure 2.11, shows oil's share of world primary energy rising, and coal's falling, before 1971–3. The 1971–3 oil price increases seem to have stabilized the share of oil; those of 1979–80 brought it down again until after the fall in price of 1986. The oil share has since been roughly stable during a period of declining prices (but other fuel prices have been declining also). There are lead times involved: the high prices in 1979–86 may have stimulated developments, for example, of nuclear power, which can compete in the post-1986 oil price world only on a variable cost basis.

Theory

Bewildered oil producers and consumers crave an explanation of why oil prices change. They need to make strategic choices to invest in and use the supply of oil or other energy sources. Within their chosen investment strategies, they have to make timing and scaling decisions. In the short term, they use actual and projected oil prices in trading and in predicting financial results.

[18] For an analysis of the capture of the European power market by cheap oil in the 1950s and 1960s see M. A. Adelman, *The World Petroleum Market* (Washington DC: Resources for the Future Inc., 1972), pp.178–83.

Figure 6.3: Primary energy: fuel shares and oil price

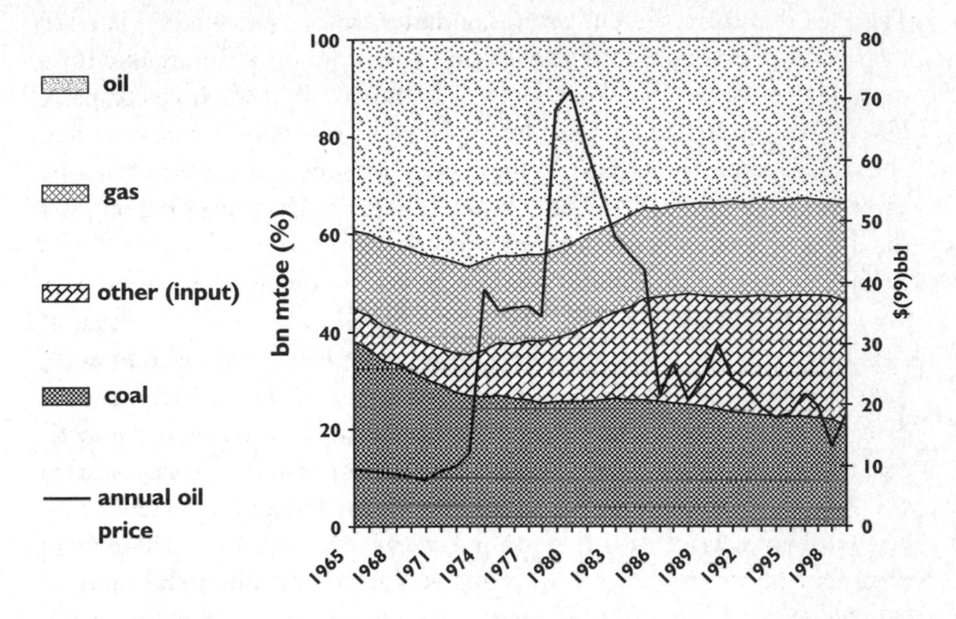

Source: BP Amoco *Statistical Review.*

Two special features of oil have attracted academic attention: the idea of exhaustible resources and the idea of there being a cartel – that is, a concentration of market power in the hands of a few producers. The puzzle is to find out how competition works in these circumstances. Different analysts and their models emphasize different aspects of the question.[19] There appear to be non-economic factors at work which limit the ability of economic models to explain or predict; but the economic factors are present nevertheless, and are present in the 'new oil economy' as well as the old.

[19] For reviews, see Djavad Salehi-Isfahani, 'Models of the Oil Market Revisited', *Journal of Energy Literature*, vol. 1, no. 1 (1995); Robert Mabro, 'OPEC Behaviour 1960–1998: A Review of the Literature', *Journal of Energy Literature*, vol. 4, no. 1 (1998).

Exhaustible resources

The idea that oil resources are finite and therefore exhaustible is most obviously true for an individual property. Production in areas with a long production history (such as the US 'Lower 48' states) has actually fallen. We described in Chapter 3 some of the main economic and physical factors which explain these declines, and the debate about the extent to which the United States can be regarded as a useful exemplar for the decline in world oil production in the future.

The key to the 'Hotelling Principle' is that an owner should compare the value (price less cost) of a barrel produced today with the value of the same barrel produced in the future (price less cost, discounted to today). The oil should be produced now if its current value is higher than the discounted value of production at a future date. In a competitive market, if all producers behave in this rational way, today's price will fall, and future prices rise, until the two are equal: the result should be that the net price will rise at the rate of discount. In mathematical form this theory can cope with changes in costs for individual producers, changes in the reserves due to new discoveries, changes in the demand for oil due to changes in taste or the development of substitutes, and different rates of discount by different producers, including different rates of discount for private sector producers and tax-collecting governments.[20] Many of these realities are difficult to observe, so that a 'Hotelling' model is difficult to test against the past. They are also difficult to predict. The 'Hotelling' idea nevertheless provides a logical basis for some intuitive reactions to broad events. The 100-year history of oil prices shown in Figure 6.2 does not show the clear upward trend which Hotelling's theory might predict. The loose story of new discoveries, nationalization, and changes in competition from other fuels meant that the variables were never fixed. Perhaps it was not that the theory was wrong, but that the facts changed.

The principle of discounting the value of future production explains some intuitive behaviour:

[20] See P. S. Dasgupta and G. M. Heal, *Economic Theory and Exhaustible Resources* (Cambridge: James Nisbet/Cambridge University Press, 1997).

- Unless prices are expected to rise, or costs are expected to fall, it is better to produce now than in the long-term future. For countries with high ratios of reserves to current production, there are powerful reasons to increase production, and this is how governments and companies have generally behaved.
- Discoveries of new reserves, or new technologies for increasing reserves from known resources, will lower the future price of oil and therefore, through discounting, lower the value of all oil in the ground. Producers should therefore increase present supply and present prices will fall. This seems, roughly, to have happened.
- Events that restrict future supplies will increase the value of oil in the ground and therefore reduce the incentive to produce today. Government actions that impose moratoria on new exploration, for example, offshore the United States, or which replace private companies (with high discount rates) by government owners (with lower discount rates) are examples of this.[21] Such changes certainly accompanied periods of rising prices.

Cartel or competition

It seems obvious that oil is neither necessarily a monopoly nor inevitably an atomistic competitive market. But what it will be in the future cannot be separated from what it 'is' today and what it has been. The ownership of the resources binds future and past structures under a common constraint. Resource ownership is determined by a combination of

[21] Morris Adelman has argued that implicit discount rates of governments in the oil-exporting countries are high because of political duties: M. A. Adelman, *The Genie Out of the Bottle* (London: MIT Press, 1995), p. 32. But this is true only sometimes. In 1974–5 and again in 1982–3, following the extraordinary price increases, many oil exporters found revenues exceeding inherited expenditure commitments; until expenditure rose, the governments accumulated financial surpluses, generally abroad. They may have implicitly discounted these for risk, or implicitly adopted as a long-term discount rate that available on foreign financial investments. Oil in the ground looked an attractive investment and governments were happy to see production fall – collectively by 45% between 1979 and 1983. Private companies (and countries such as Mexico, China and Brazil), which were not in this position, continued to get their oil out as fast as possible.

geology and government. Just over 80% of the world's proved resources occur in 30 reservoirs. Eleven of them, accounting for 70% of the world's proved reserves, are in the Middle East, where they are owned by the states concerned.[22]

This distribution makes it difficult to guess at the counterfactual. What would have happened to oil production and prices if:

- most of the world's oil reserves had been distributed among a large number of private sector corporations, as they are in North America;[23]
- there were no government intervention in investment or the rates of production; and
- access to new resources was open to competition (so that owners of 'old' expensive oil could not monopolize the availability of 'new' cheap oil)?

If such a situation had existed in the past, we can guess that the development of low-cost reserves in the Middle East, Africa and Venezuela would have been more rapid. The US oil industry would have closed down (as the British coal industry has) in the face of cheap imports. Low prices would have created an even greater demand for oil. The gloomy predictions of geologists and petroleum engineers in the late 1960s would have been fulfilled: the expansion of 'cheap' oil would have been unsustainable. Over a 20- or 30-year period we might indeed have seen prices rising on a trend of the type 'Hotelling' might have predicted, but from a starting point that would have been lower in 1970 than it in fact was. A test of threats to the 'Hotelling' equilibrium is to compare the ratios of reserves to production in different countries. For any given path of oil prices, cheap oil should be developed first and expensive oil later. There would be no incentive to add to reserves

---

[22] T. R. Klett, T. S. Ahlbrandt, J. W. Schmoker and G. L. Dolton, *Ranking of World's Known Oil and Gas Provinces by Known Petroleum Volumes*, Open File Report 97-463 (Denver: US Geological Survey, 1997).

[23] In both the United States and Canada offshore resources, and resources under federal, state or provincially owned land, are the property of the state concerned, which leases the right to discover and develop these reserves to the private sector.

in high-cost countries. The ratio of reserves to production would tend to be equal in all producing countries.

The facts, however, have been very different.

The imbalance of reserves

The 'expensive oil' was discovered first.[24] Up to the 1970s the 'cheap' oil was being discovered in quantities far in excess of what was needed to cover demand. Geologists from private sector companies competed to find big fields because their costs would be lower: the company with the highest proportion of low-cost fields would grow faster and more profitably than companies dependent on high-cost fields. The sequence of discoveries created a huge disequilibrium. Its effects were mitigated by various factors. US production was partly sheltered by import controls from 1952 to 1971 (effectively at equivalent to a tariff of $1–2/bbl; around 50% in the prices of that time). Most development outside North America, the Soviet Union, China and Mexico was in the hands of (relatively few) private sector companies, most of which also had an interest in protecting their investments in 'expensive' oil in North America. Their investment and production plans were in any case dependent on negotiation with the 'host' governments from whom they held concessions (see above). These governments in general pressed for rapid development of newly discovered reserves. Nevertheless, the ratio of reserves to production still varies widely from one country to another. At the end of 1998 the ratio was 83 in the Middle East, 10 in the United States.[25] This also remains the motor for future competition within the Middle East, discussed later in this chapter.

Competition among the companies was reinforced by the fact that they suffered differently from the nationalization and expropriations in the OPEC countries. Governments in non-OPEC countries were concerned with the stability of future oil supplies. Importing countries

---

[24] In a theoretical 'Hotelling' world, oil would not be discovered until it was wanted, and ratios of reserves to production in different countries would converge at the level necessary to cover the lead time between discovery and production: the 'working inventory'. In the United States this seems to be about 10 years.

[25] *BP Statistical Reviews*, 1972 and 1998.

competed to attract company investment in new oil and gas provinces as these were discovered, for example, in the North Sea. There were high cash flows in companies in the period 1973–83 to fund exploration and development; these may have been misguided in terms of price expectations, but they increased reserves outside OPEC and the Middle East. The experience created an infrastructure of capital and knowledge in the private sector which gave it the capability to reduce costs after the oil price collapse of 1985–6. As a result, the private sector has continued to add 'high-cost' reserves both by new discoveries – for example, offshore – and by 'growing' reserves through increasing recoveries (see Chapter 2).

The 'high' prices of 1975–85 had similar effects in nuclear energy, coal and natural gas. The suddenness of the price changes and the publicity given to the uncertainties surrounding them created a policy and commercial logic for improving the efficiency of using fuel and industries to provide the means to do it.

Cartel chimera

The theoretical studies tend to focus on whether a cartel can survive. The theoretical answer is that 'cheating' can be punished and a cartel need not break down if the main producers are in the game for the long term. For the oil industry, geology ensures this: resources are where prehistory put them and cannot be manufactured by new entrants.[26]

Before 1973, negotiations between international oil companies and governments – sometimes collectively – established prices to be used for tax purposes which established effectively a floor price for most oil in international trade.[27] Since the 1973 OPEC decision to set prices without negotiation with the private sector companies, attention has focused on OPEC or a core of producers within OPEC as a potential 'cartel', with non-OPEC producers forming a 'competitive fringe' of price takers which do not seek to affect the international price by their

[26] Salehi-Isfahani, 'Models of the Oil Market Revisited', p. 12.

[27] Gavin Brown, *OPEC and the World Petroleum Market* (London: Longman, 1986; 2nd edn 1991).

production or investment behaviour.[28] The 'core' comprises countries where the oil industry was under state ownership and control; the 'fringe' includes all of the private sector, countries with mixed private–public sector ownership and control, and non-OPEC state producers. Since the 'fringe' are supposed to be price takers, the effect of unexpected changes in the supply–demand balance falls disproportionately upon the 'core'. With a market share of 30% in 1985, the ability of OPEC to act as an effective 'cartel' was accordingly limited; even with a market share of 42% in 1998, the 'OPEC cartel' still failed to prevent another collapse in the oil price.

The difficulty of working out these theories on the historical record for oil is that so many of the parameters changed. Countries expanded their spending at different rates: the Iranian Revolution permanently reduced Iran's current production capacity and certainly delayed major investment in new capacity; the Iran–Iraq War absorbed resources of both countries which might otherwise have been used either in expanding oil and gas capacity or in diversifying their economies (so as to reduce the role of oil in future policy); and the Gulf War again dealt a double blow – at Kuwait's production in the short term and at Iraq's expansion during the 1990s.

It may be impossible for the core Middle East countries to agree on a 'fair share' of output. There is an imbalance within the cartel core between high-reserve and high-production countries, similar to that in the world at large. There are big differences in the ratios between reserves and production, although the underlying cost structures are not very different (though it is true that Iran, with the lowest reserve ratio, probably has the highest incremental development costs). There are imbalances between production capacity and production: Saudi Arabia has higher capacity in its shares of reserves and of production (which are more or less in line); Iran's position is the reverse. Kuwait and the UAE, given their declared reserves, are underdeveloped, but also have the smallest shares (excepting Qatar) of the group's population.

It is difficult to construct theories about the effect of population growth on the governments' objectives. One idea is that an exporting

[28] See Mabro, 'OPEC Behaviour 1960–1998'.

Table 6.1: Percentage of the cartel core assets

	Production 1999 %	Capacity 2000 %	Reserves %	Reserve/production ratio 1999	Population %
Saudi Arabia	42	47	40	92	18
Iran	19	16	14	70	56
Kuwait	11	12	15	136	2
UAE	11	11	15	129	3
Quatar	3	3	1	16	1
Iraq	14	12	17	122	20
	100	100	100		100

Sources: BP Statistical Review, IEA,DOE (EIA 00)

government–owner will set output or prices so as to satisfy a current revenue target which will reflect the 'needs' or expectations of the population. This target will move over time as the population increases and expectations change.[29] The target creates, in the short term, a 'backward' bending supply curve: if prices rise for external reasons, as in 1979–80, some producing governments may be happy to see production fall because revenue surpluses would arise from higher production. The financial surplus would accrue in foreign currency in the US financial markets, and would not be valued as highly as oil in the exporter's own ground. The US freezing of Iranian assets in the early 1980s underlined the risk of oil producers making financial investments in the United States. Unilateral falls in 'surplus' output help support the price but do not need to be coordinated. Eventually, however, this 'willing withdrawal' of supply may be insufficient to contain the fall in demand (as in 1983–6), and all exporters face a situation in which they are unwilling to withhold supply, competition resumes and prices fall.

When prices are falling to low levels the situation is different for governments in countries where oil exports are the major source of foreign exchange and government revenue. The current revenue 'needs' are urgent: they are a 'survival level' for the government similar to subsistence

[29] Population growth rates in the Middle East oil-exporting countries have in the past 20 years been among the highest in the world, in some cases partly due to the expansion of a population of migrant labourers who, however, do not enjoy the same access to government revenues as hereditary nationals.

wages for the survival of labour or debt service for the survival of an enterprise in the private sector. If the 'survival levels' are not covered the entity will cease to exist – even if the revenues exceed the technical costs of production. Under such conditions, the core oil-exporting countries have shown that they can agree to stop competing and allocate production by quotas.

Collective action to restrain output more or less stabilized the price in 1986 and again in 1999, though in 1999 cooperation from Norway and Mexico, as well as the non-core members of OPEC such as Venezuela, was essential to the result. In both cases a key to agreement was to base restraints on current actual production, rather than on theoretical calculations of what quotas 'should' be, or on the last negotiated set of quotas. Moreover, agreement was difficult to reach: the market shares had changed considerably, and the likelihood that quotas would be based on market shares meant that competition immediately before the agreement tended to intensify, driving the price further down.

Inevitability of competition

For importers, the good news is that oil exporters inevitably restart competition when demand increases and prices rise. If quotas were maintained absolutely, prices would continue to rise for a time (as in 1990). As revenues rise, those exporters capable of increasing production from existing capacity want to expand production. They may 'cheat' by exceeding their quotas. If the exporters try to limit competition, quotas have to be renegotiated to take account of the factors which could be ignored when all were in 'survival' mode, which is very difficult. Either way, market shares change as a result. There is no escape from this. In the short term, exporters almost never have the same shares of short-term spare capacity as they have of production.

The difficulty of negotiating changes in quotas to match changes in demand can, in theory, be revised by the idea of a price bracket: quotas apply when the price is at a floor, and are lifted when it moves away from the floor by a prescribed amount. This appeared to be the concept of the price bracket floated at the March 2000 OPEC meeting, but not embodied in a formal OPEC decision. The problem with the 2000 price bracket concept may have been that a wide bracket was set, starting at $24 – well above the 'floor' of $10 at which quotas were imposed, and above the average of prices for the previous five or more years. The prospective increase in supply at the top of the bracket was rather small (500,000 b/d when the price exceeded $28/bbl). Moreover, when the price exceeded the upper trigger at the beginning of June the automatic increase did not materialize: decisions were deferred and the price continued to rise. Agreeing to relax quotas in response to a clear price signal would be a more honourable way of relaxing quotas than cheating and easier than renegotiation. If a sensibly defined price bracket were actually used semi-automatically, attempted production might respond more quickly to changed market conditions. But either way, the competition is bound to resume among exporters with different endowments of capacity and reserves and different 'needs' for additional revenue.

In the longer term, the cartel core – or OPEC as a whole – has found no way of negotiating agreement on capacity expansion.[30] Current capacity is not held in the same proportion as reserves (and present reserves may not be proportionate to potential future resources). Are governments likely to agree on an allocation of expansion rights?

Agreement on a long-term price target might be a way of allowing competition while signalling some probable willingness to cooperate in future to restrain production if targets were not met. But even if long-term price targets were set by the core countries – or even by OPEC as a whole – the uncertainties of demand and non-OPEC supply would be reflected in fluctuating demand for supplies from the cartel core. Each government exporter would have to decide what capacity to build to capture market share (from those with limited capacity expansion) when demand, and therefore prices, were high. There would be a tendency to overbuild. If there were some pretence that the long-term target was set to preserve the market share of oil against competing fuels, there would be an additional reason for overbuilding to provide spare capacity for use when prices exceeded the target.

[30] For an account of OPEC's attempt to allocate expansion pro rata in 1965–6 see Brown, *OPEC and the World Petroleum Market*, A.4.12.ii.

Figure 6.4: Covariance of price and market share changes

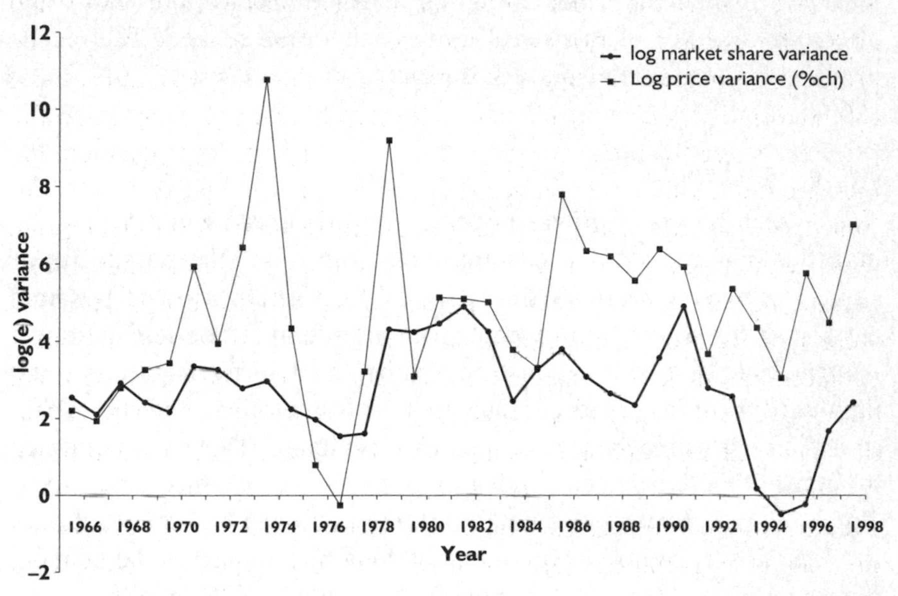

Note: The covariance of the two indices is 96%.

This suggests that stop–start competitive behaviour by the government oil exporters is inevitable. Competition may 'stop' when prices are suddenly pushed so high (as after the price shock of 1979–80) that a number of producers independently decide to keep oil in the ground rather than convert it into financial assets in the United States. This situation does not last, because spending habits adjust to higher incomes and the choice disappears. Competition may also be temporarily 'stopped' when it has driven prices so low that all major exporters see revenues at survival levels. Competition between the major exporters is, however, driven by fundamental differences which cannot be suppressed by intergovernmental agreement, while shares of production are not proportional to shares of capacity and shares of reserves. Markets will allocate shares while competition is not in the 'stop' mode – which is likely to be most of the time.

If prices go up for cyclical reasons the exporters are capable of expanding output in the short run to different extents, because of their different investments, which in turn reflect their different potential. Different investments and changing shares of output can also drive prices down. Price changes and share changes are related. The record of this relationship is displayed in Figure 6.4.

Institutional context

The concentration of oil resources and government control over the industry from the main exporting areas both raises the possibility of cartelization and destroys that possibility. Cartelization is possible because governments are exempt from anti-competition legislation; it is impossible to sustain because governments are differently endowed – they will not trade these endowments with one other, and the differences in endowment make competition inevitable. They will not agree to stop competing except under extreme and temporary conditions. Competition between government oil exporters can be prevented only if all the governments concerned agree to deny themselves the benefits of competition when it seems beneficial to some of them. There is no higher authority.

The model of the Texas Railroad Commission is commercially apt but institutionally inappropriate. In Texas in the early part of the twentieth century, competition between competing private oil and gas producers was controlled by a Railroad Commission of public officials, who could bring some public interest into the decision and impose it by regulation backed by the authority of the state government. There is no supranational authority for oil, nor is it likely that one will be tried – along the lines, for example, of the High Authority for Steel in the European Coal and Steel Community after the Second World War. A UN Commodity Agreement of the type envisaged in Article XVI of GATT is theoretically possible.[31] There already exist such examples as the International Agreements on Wheat, Coffee, Tin and Cocoa, all of which, though they vary considerably, do establish treaty commitments for various types of action on quotas by exporters when prices pass certain threshold levels. Importing countries are involved in the negotiation of the

[31] Adelman, *The World Petroleum Market*, pp. 247–9.

price triggers and reinforce the power of quota restriction by undertaking to limit imports from non-members when prices are low. The renegotiation of prices and quotas at three-year (or similar) intervals allows some scope for market shares to change as a result of normal commercial and investment activity so long as prices are above the specified floor. The record of such agreements in their objective of stabilizing prices is mixed. The concept of multilateral government intervention in a commodity market runs against the prevailing faith in the benefits of free trade and investment.

Political rivalry

The governments of the principal oil exporters are not politically neutral in their relations with each other or with their markets. Venezuela has little political common ground with the Middle East oil producers except on oil matters. Much of even this common ground is now history: in OPEC's first 20 years its members were concerned with common demands on the major oil companies operating in most of their countries, and since these foreign operations were replaced by government or government-controlled state companies the questions of common attitudes towards tax reference prices, royalties, international arbitration and the like have disappeared from the OPEC agenda. Venezuela's re-introduction of foreign oil companies to its oil industry since 1991 has been on its own terms and has not been coordinated – or even discussed – with other OPEC countries. Outside oil, and a very general community of interest as developing countries, Venezuela's geopolitical context is quite different from that of the Middle East oil exporters. Venezuela is a western hemisphere country, and a member of hemispheric and regional organizations which include the United States and other oil-importing states. Other than concern with national independence and economic welfare, there is no structural opposition of interest between Venezuela and the United States. Venezuela also subscribes to, and tends to implement, the values which the United States espouses of democracy, human rights and increasingly liberal international rules for trade and investment.

The Middle Eastern oil exporters are in different positions. As well as concerns about national independence and underdevelopment, they share hostility to the position of Israel and the US connections with Israel which are an essential part of US policy in the region. During the Cold War period Iraq developed connections with the Soviet Union, and, more recently, Iran has looked to Russia as a source of military and nuclear technology. Iran and Iraq, were it not for the respective US and UN sanctions, would have leverage over the development of independent export and trade routes for the Caspian successor states of the Soviet Union. The Middle Eastern oil exporters can also look to developing relations with East Asia as a source of investment as well as a market for oil independently of the United States and Europe: already more than 60% of the region's oil exports go east, not west. The states relate differently to the rest of the world in general and to the US in particular. Iraq since 1990 has been the subject of UN sanctions enforced militarily by the United States and some of its allies. Iran is subject to a variety of sanctions, some of which (under the Iran–Libya Sanctions Act of 1996) apply to US allies to prevent their developing normal economic relations with Iran. At the other extreme, Kuwait and the UAE have formal security agreements with the United States. Saudi Arabia, which is not party to such an agreement, nevertheless received US protection during the Gulf War, is a major purchaser of US arms and related services, and is technically positioned to receive US military protection again. Participation by the Middle East oil exporters in multilateral institutions and conventions is also kaleidoscopic. Iran is a member of most multilateral agencies and a signatory of the major covenants on human rights. Saudi Arabia and the UAE, like Iraq, are not signatories of the main human rights conventions, nor are they members of GATT (though Saudi Arabia has now applied): both countries are, however, members of the IMF.

Saudi Arabia has been perhaps the most adept of the Middle Eastern oil exporters in maintaining a continuously broad and deep set of economic and security relations with the United States and with the multilateral institutions while maintaining profound ideological opposition to the existence of Israel and US support for it. At the same time Saudi Arabia has avoided implementing in its own territory the institutions which the United States and other countries regard as essential to the values of democracy and personal freedom. This double act (cooperating with the wider world on economic and security matters while rejecting its influence on domestic affairs) has given Saudi Arabia (and, in a less contradictory way, the other GCC states) great advantages over other countries like Iran and Iraq in terms of security, access to finance and technology, and a claim on the attention of policy-makers in the world's major industrial countries.

Differences in the Middle Eastern states' relations with the rest of the world are paralleled by differences among the states themselves. They are rivals in lifestyles. For a region with a long history of independent states and political organization, never colonized by European industrial powers, the diversity of political systems is remarkable. The major states have governmental systems which differ greatly from one another, but none of which has serious parallels outside the region. Iran is an Islamic, pluralist state with a written constitution, an elaborate set of working checks and balances in its government, and recent experience of changing parts of its leadership through parliamentary and mayoral elections. It also has centuries of history as an independent state. Iraq is an absolute dictatorship. Outside Iran and Iraq, the Gulf contains most of the world's few remaining monarchical rulers. Kuwait has a parliament with powers analogous to the legislature of many pre-independence British colonies in the 1950s. Saudi Arabia and the other Gulf states have essentially monarchical governments in which the Islamic legal system provides a degree of judicial independence but economic matters are not brought to public account in ways familiar in the OECD. Although all are Islamic, the dominant tradition differs from country to country, reflecting not only the split between Sh'ia (dominant in Iran) and Sunni (dominant elsewhere) but between Wahabi (dominant in Saudi Arabia) and less fundamental traditions. There is also an immense gap between the essentially conservative Wahabi approach to the modern world and that of more radical movements.

All of these differences add up to a context of geopolitical and regional rivalry. The Arab oil embargo of 1973, provoked by US military assistance to Israel in the face of Egyptian attack, was a truly exceptional political event. The enduring geopolitical, constitutional, civil and religious structures of the Middle Eastern oil-exporting countries are those of rivalry, which supports competition for revenues and the power that they bring. The context is not one which supports the acceptance of restraints in the interests of a higher common good.

Dominant player

Saudi Arabia's position in the extraordinary rivalries of the Middle East is a permanent reason for its government to be cautious about exercising the market power for which it is qualified by its command of the largest single concentration of low-cost reserves in the world. The larger populations and armies of Iraq and Iran constitute a continuous potential threat of adventurism: Saudi territory was invaded by Iraq in 1990, and during the preceding decades Iran threatened traffic in the Gulf. Saudi Arabia and Iran compete for influence in other Islamic states in the region and in Central Asia. The Saudi Arabian government needs oil export revenues to buy sophisticated military equipment to compensate for its disadvantage in population numbers. It needs not only foreign suppliers but foreign technical support to keep the equipment going, and needs to behave in a way which will ensure that a global power or powers will protect it because of its oil supply. High volumes and low prices could achieve this, but could also destabilize and provoke the neighbouring rival countries.

If Saudi Arabia and its reserves were situated outside the Middle East its government could have less consideration for other exporters and pay less attention to importers. As it is, and where it is, it will inevitably be motivated to avoid the sudden development of 'high' or 'low' prices – though this may not always be achievable.

All cartel core or dominant firm theories for the oil market feature a 'multiplier'. This means that all changes in the balance of supply and demand fall to be accommodated in the end by the Kingdom of Saudi Arabia. The Saudi choice is how far to accept, and how far to force, the sharing of the accommodation with other 'core' exporters or even (in 1999) with exporters outside OPEC. There is, however, a difference between balancing short-term disruptions (Saudi Arabia acting as emergency supplier to soften price shocks to importers) and balancing medium-term disequilibria (Saudi Arabia acting as 'swing producer' because capacity expansion elsewhere has run ahead of demand).

The record shows attempts at market management: stable shares meant stable prices, but changes could be led by either factor. Rising prices before 1973 went with a rising Saudi share of world production (Saudi is the country with the biggest potential to expand production). After each shock to other Middle Eastern suppliers the Saudi share (the emergency supplier for short-term disruption) rose further; but after each price spike, as other supplies came forward and prices fell, the Saudi share of Middle East supplies fell towards its previous level. Between 1983 and 1986 Saudi Arabia accommodated more than its share of the fall in demand (the 'swing producer'), presumably with some effect on price and the revenues of other exporters. In 1985 the Saudi policy changed: it decided to defend its market share, and was able to induce sufficient restraint in other exporters to stabilize the price or at least

Figure 6.5: Saudi Arabia's balancing supply roles

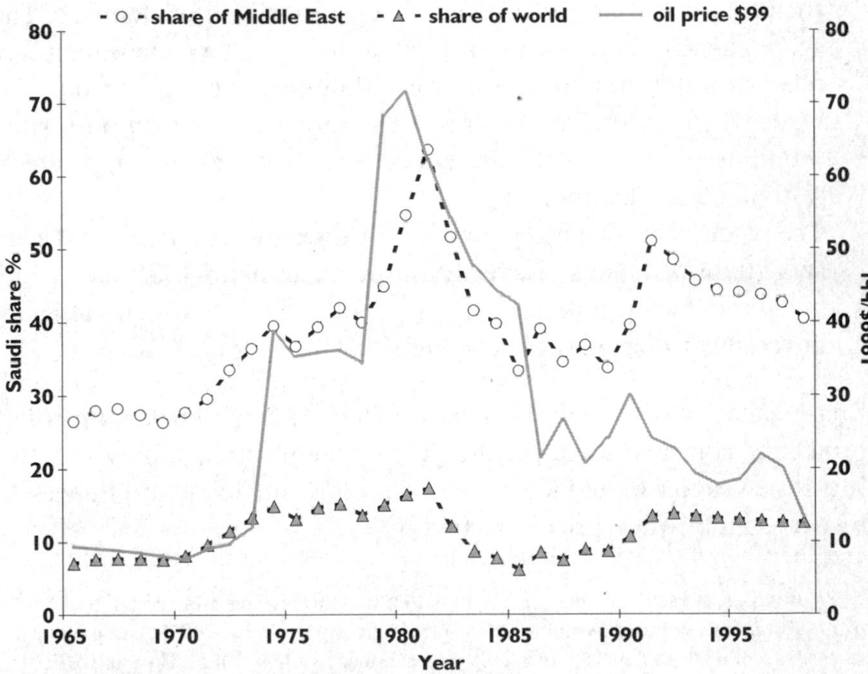

Source: BP Amoco *Statistical Review*.

limit its decline.[32] In 1990–1 Saudi Arabia (again as 'emergency supplier') increased supply and gained market share. Its market share eroded slowly as Kuwait's and Iraq's supplies were restored, but by 1998 it was around 40% of the Middle East supply: better than in the late 1980s, but no better than in the mid-1970s. The Saudi share of world production was lower because the Middle East share was lower.

Saudi Arabia's exposure, and also its strength, in its role as emergency supplier arises from a combination of several factors:

- The oil resource is wholly owned and (since 1976) totally controlled by the Saudi government; there are no private interests to accommodate. Because of the prolific nature of the few large reservoirs, the labour and multiplier effects of variations in production are very small.
- More than any other exporter, Saudi Arabia has the oil reserves and technical capacity to expand production capacity rapidly and cheaply; the financial strength of the kingdom so far has been such that it can afford to expand capacity ahead of demand. The spare capacity enables Saudi Arabia, to a greater extent than any other exporter, to compensate for falling prices by increasing volumes when prices are too 'low'. The threat of such action (or the action itself, as in 1998) can induce other exporters to cooperate in restraining production.
- The same spare capacity provides a strategic reserve to take up short-term or 'shock' increases in demand, as in 1979 and 1990, with the effect of mitigating rises in price to levels which the Saudi government judges to be 'too high'.

Spare capacity thus underpins both Saudi Arabia's leadership roles, protecting consumers against the persistence of price spikes and giving it a weapon to induce restraint among sufficient producers to restrict supply when prices are falling.

[32] 'With half its capacity still unused, the kingdom was earning an estimated $43bn a day from substantially increased volume of low-priced oil imports in June 1986, compared with its earnings of £38bn a day from officially priced imports in June 1985': Brown, *OPEC and the World Petroleum Market*, p. 335.

Figure 6.6: Saudi Arabia: absorption of oil revenue

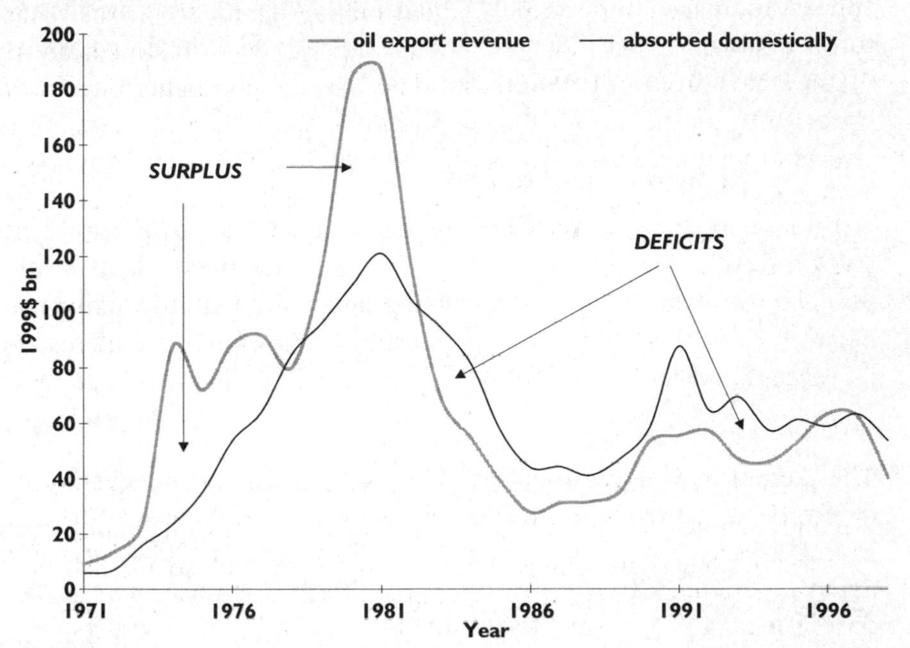

Note: 'Absorbed domestically' in the graph is the difference between oil export revenue and the sum of line 78 *bjd* (balance of direct, portfolio and other investments abroad) and line 79 *dad* (changes in reserves and related assets: reserves, exceptional financing, IMF credits and loans).

Source: IMF, *International Financial Statistics Yearbook 1999* (Washington DC, 1999), Vol. LII, pp. 456–7.

There has been one important change in the Saudi Arabian position since 1979. The kingdom generated huge balance of payments surpluses between 1974 and 1982, but the rate at which it absorbed export revenue grew dramatically, so that since the oil price collapse of 1986 it has, in most years, used more foreign currency than its oil exports generated. It has become a 'high absorber' of revenue. It is no longer the case that additional revenues may be undervalued because their only use is to be converted in to foreign financial assets. Figure 6.6 illustrates the point.

After the shock price increases of 1973 and 1979 oil revenues exceeded even the 18-fold increase in domestic absorption. Domestic use was dragged down in response to falling revenue but never below it, except in 1996. The kingdom is probably still a net creditor, and appears to have stabilized its absorption by budget and economic reforms since 1990, but it cannot be indifferent to additional revenue as it could have been in previous decades.[33]

Prices in the 'new economy of oil'

To guess at the future price of oil requires an understanding of its present level, a view of the medium-term cycles around the trend which is suspected,[34] and consideration of whether a new episode in the drama has begun.

'The present': 1998–2000

The present is difficult to define. The spot price of the benchmark of dated Brent fell from around $21 in October 1997 to around $12 in March 1999,[35] rose fairly steadily to $30 in March 2000, and fell again to around $20 in April. By June 2000 it was once more above $30. Three- and nine-month futures prices did not follow these extremes. It is arguable that these fluctuations in the spot price were more extreme than would have been the case in an atomistic market. Hesitation in the face of falling Asian demand and a reluctance to accommodate rising Iraq production within OPEC led to a mistaken OPEC decision to increase quotas in November 1998. The difficulties of renegotiation combined with uncertainty delayed increases in

---

[33] There should still be a positive contribution of $132bn to the kingdom's external balance sheet, assuming all the investments have at least maintained their real value.

[34] The problem of the wrong baseline persistently bedevils understanding. Applying the 'Hotelling' rule to the 1980 prices, as many modellers did, assumes that 1980 prices were the result of a rational and all-seeing process. See Hillard Huntingdon, 'Oil Price Forecasting in the 1980s: What Went Wrong' *Energy Journal*, vol. 15, no. 2 (1994), pp. 1–22.

[35] Crude oil is not homogeneous: oil from different reservoirs has different physical and chemical properties and therefore different values and prices. For an explanation of why the oil produced from the Brent reservoir in the UK North Sea has become a benchmark, see Horsnell and Mabro, *Oil Markets and Prices*. For analyses of the price movement between October 1997 and March 1999, see Robert Mabro, *The Oil Price Crisis of 1998*, SP10 (Oxford: Oxford Institute of Energy Studies, 1999) and IEA, Background Paper for Seminar on the Effects of the Oil Price Drop of 1997–98, 14 May 1998.

supply in 2000. It is difficult to believe that a price of $30 can be sustained into the medium or longer term for reasons given below.

What have we learnt about defensive restraints on competition?

- Deciding when and how to restrain competition has again proved very difficult, even in a desperate situation for the exporters. As on previous occasions, OPEC has introduced restraints on competition only in the face of a free fall in oil prices.
- The restraints were allocated broadly on the basis of production levels immediately before the collapse, not on the basis of earlier quotas or any formula.
- This time, the cooperation of some non-OPEC producing governments (Mexico, Norway and Russia) was required to achieve the necessary restraint.
- The only way (in April 2000) OPEC could find of agreeing on increases in quotas was to withdraw the last round of cuts.[36]

The participation of non-OPEC producers was the major change to previous behaviour, broadening the base over which production restraint was exercised. Price defence in future is likely to be more effective than at any previous time (non-OPEC exporters were not involved, by definition, in the formation of OPEC as a price defence organization, or in the quota system of the 1980s). Non-OPEC participation makes absolutely clear that the pricing policy has a defensive, economic basis, rather than the political motivation often attributed to it. It is this last feature in particular that is likely to mark the beginning of a new era: the threat of price 'aggression' which underlay importers' fears about security of supply before 1985 is diminished by the participation of two OECD members in the price defence arrangements.

The reasons for the collapse in the price of crude oil have been extensively discussed, though little foreseen.[37] With different emphases, most commentators point to:[38]

---

[36] For details, see IEA, *Oil Market Report*, 11 April 2000, p. 14.

[37] Ian Seymour and Robert Mabro, 'Oil Price Outlook Through 2000', *Oxford Energy Forum*, no. 24, February 1996.

[38] See e.g. IEA, Background Paper for Seminar on the Effects of the Oil Price Drop of 1997–98, 14 May 1998; Mabro, *The Oil Price Crisis of 1998*.

- A mild winter in several parts of the northern hemisphere, depressing heating oil demand.
- The resolution of a number of difficulties which had delayed the start of the Iraq oil-for-food export programme. Iraqi oil exports increased by nearly 1.5m b/d during 1998. Because the limit to the oil-for-food programme was defined in money, not volume terms, Iraq was able to increase the volume of its exports as the price fell.
- The developing financial crisis in some Asian countries. This led to expectations of demand in Japan, Korea, Thailand and Indonesia being revised from a growth of about 0.5m b/d to a fall of about 0.5m b/d, with continuing fear that Chinese demand would follow a similar pattern.
- A decision in December 1997 by OPEC to raise its normal production ceiling from 25m b/d to 27.5m b/d, regardless of all the adverse developments in demand. This largely accommodated increases in production which had already taken place in Venezuela in disregard of the previous quota ceiling. With hindsight, the actual increase in supply resulting from this decision may have been less than 0.5m b/d, but the psychological effect was very great.[39]

All of these were transitory events. The unusual demand circumstances came to an end: growth in Asian demand resumed; the contango in the price disappeared. Production restraints were re-established, in a stronger form than before: they were based on prior actual production rather than the theoretical quotas, which had got out of line with actual production. The degree of compliance with the cuts rose steadily, reaching over 90% in August 1999.[40]

The 1999 restraints were marked by three enduring implications:

- There was an easing of competition between Saudi Arabia and Venezuela, with the effect that Venezuela may in future respect quotas. Competition between these two major suppliers was therefore stabilized. This development was underscored by a change of government

[39] Robert Mabro, using non-IEA data, argues that the effects of inventory build-up and production increases at the end of 1997 were less than suggested by the IEA, but that the appearance of a contango in the market nevertheless signalled excess prompt supplies.

[40] IEA, *Oil Market Report*, 10 September 1999, p. 18.

in Venezuela and a change of management in Petroleos de Venezuela, with that organization brought under close government control as far as oil production and investment policy are concerned.

- Non-OPEC producers joined the restraint: in particular, Mexico (which brokered the original reconciliation between Saudi Arabia and Venezuela early in 1998) and Norway, which for the first time announced and sustained significant production cuts as part of a concerted price defence by exporters. Russia and Oman also contributed.
- There has been a *rapprochement* at the political level between Saudi Arabia and Iran, based on the Iranian President's visit to Riyadh early in 1999.

The overall effect was to restore the price to its previous level by mid-1999; to resolve, for the time being, two of the serious internal conflicts between major OPEC exporters; and to bring major non-OPEC importers into the price stabilization arrangements. The questions then arose how rapidly importers would rebuild stocks, how rapidly Asian demand would recover, and what would be the delayed effect on non-OPEC production of the spending cuts triggered by the price collapse. The situation was complicated by shortages of gasoline in some US states for reasons not connected with crude oil supply (tighter gasoline specification coincided with a reversal of the earlier policy to allow and encourage the use of methyl-tertiary-butyl ether (MTBE; not derived from crude oil) to meet the specifications. As always, uncertainties about actual production, demand and inventory data made it difficult to read trends. OPEC's problem was that it was not, like a company management, in daily session, capable of correcting mistakes quickly. Scheduled OPEC meetings took place every three months; unscheduled meetings would disturb the market; secret meetings were virtually impossible. The 'automatic' production mechanism to add 500,000 bbl/d to production when the average price of a basket of crudes exceeded $28 for 20 days was ignored when this happened early in June. The exporters did not all have the capacity to take advantage of rising demand:[41] Iran, in this situation, declined to agree to the new targets.

[41] For an assessment, see IEA, *Oil Market Report*, 11 April 2000, p. 14.

The medium term: another cycle or another path

Growth in demand over, say, 2000–5 will be dominated by economic activity, which (in 2000) presents a mixed outlook. There are international imbalances which seem to call for correction. At the heart of them is the continuing US propensity to expand private consumption and the failure of private consumption to expand in Japan. Other issues are also important: whether the US economy is really in a 'new paradigm' of higher sustainable long-run economic growth; whether Europe is facing a cyclical upswing; and whether the oil price rise of 1999–2000 will affect inflation sufficiently to shift macroeconomic policy in the importing countries.

On the supply side, the success of exporters in defending the floor may not be matched by success in raising it or in lifting the price towards the limit set by medium- and longer-term expectations. The private sector companies scaled back capital investment plans – by about 30% – in 1999 in response to the drop in revenues in 1998. Priority seems to have been given to new offshore opportunities in the Gulf of Mexico, offshore West Africa and offshore Brazil. There has also been a sustained interest by the major international companies in expanding their portfolios of natural gas (see Chapter 9). In the government-controlled sector, government intentions for investment are not clear. Before the price recovery and rise in 2000, there were negative tendencies, which persist at the time of writing:

- In Venezuela, following the change of government and regime in Petroleos de Venezuela (PDVSA), expansion plans were curbed and no new licensing rounds or Association ventures seem likely in the near future.
- In Russia, the production-sharing agreement (PSA) projects remain stalled, except in Sakhalin; the foreign equity investments in Lukoil, Sidanco and other enterprises do not offer an immediately attractive route in the light of the difficulties of 1999–2000 in the control and accountability of Sidanco, a company in which BP had a significant interest.
- In Kuwait, terms for significant foreign investment have not yet formally emerged, despite five years of discussion.

- Foreign investment opportunities in Saudi Arabia have in fact been limited to opportunities to invest in gas for the local petrochemicals industry.
- In Iraq, government willingness to open the sector to large-scale and rapid development with the international private sector continued to be frustrated by UN sanctions.

On balance, it is likely that oil demand will continue to rise (mainly with Asian growth) through to 2005. But it is unlikely that new surplus capacity will be developed over the next three to five years unless the Saudi Arabian government decides to implement a further expansion as a matter of policy, designed to support and strengthen its role as a market leader. In this scenario a strengthened role for Saudi Arabia might be some substitute for the unity of purpose and flexibility of action necessary to keep the oil price cruising above the floor.

The longer term: the 'new economy' of oil

For the long term, the constraints on oil price arise from outside the oil industry. The number of different sources of constraint, and their strength, define a new arena for oil. It is now a commodity facing competition from without as well as within. The temporary high oil prices create opportunities which will be taken by competing fuels and technologies as well as by new oil developments.

In the 'conventional vision' described in Chapter 2, competition up to 2020 arises mainly from the increasing share of non-conventional oil in the active reserve base and from the increasing substitution of natural gas for oil in power and other markets around the world.

Chapter 3 suggests that competition to bring oil resources to the market will continue, but the geographic reach will widen to include sources of non-conventional oil and the technical reach will place ever more emphasis on increasing production from known resources. Expectations about long-term reserves and resources seem unlikely to change as dramatically as in the past. New oil reservoirs which will have the same proportional impact on reserves as the discoveries of the 1950s and 1960s are unlikely to be discovered in the next 20 years.

Expectations will be influenced less by discovery than by technical changes which will have an incremental effect across known reserves and resources. Even if some of these – for example, to do with downhole stimulation of heavy oil – have a large potential effect, the impact on expectations will be less, say, than the discovery of a new province of conventional super-giant reservoirs. For owners of existing resources, the calculation of whether to 'develop now or develop later' is less likely to be upset by changes in the perception of resources.

Chapter 4 suggests that even in the transport sector, oil, and the makers of vehicles designed to use oil as the principal fuel, will face 'competition from without'. There will be competition to develop new liquid fuels for conventional engines; competition from technologies like fuel cells for vehicle propulsion; and competition among transport systems to meet the environmental and social demands flowing from restrictions on space, air pollution and noise in major cities.

Chapter 5 described the uncertainties that exist about the strength and timing of gas substitution and the market dynamics which may link it to oil. The continuing liberalization of electricity and gas markets, the development of clean coal technologies, and investment in cross-border gas infrastructure will all tend to increase competition against oil, either directly in the power sector or indirectly in markets where oil competes with electricity. As gas-to-gas competition develops in Europe, Central Asia and the Middle East, the big picture for gas will look more like oil in the 1960s: owners of gas resources will compete to get their reserves developed early rather than late.

Long-term price: old band, new strength

A figure of around $10 has seemed to be the threshold level in the short term because it provokes OPEC market power temporarily to freeze competition among exporters. Below that figure some high-cost non-OPEC producer might begin to see little margins in their variable costs: workovers and infill drilling would be slowed onshore (these adjustments are less relevant to offshore producers with high fixed costs and different production strategies). The same argument might apply to competing natural gas supplies in switchable markets such as power grids. In the medium term such prices could be sustained against demand trends expected in the 'conventional vision' (augmented by the effect of price on demand) only by rapid expansion of low-cost production in the Middle East, Venezuela and Mexico. New offshore developments would be slowed down, as would competing gas developments.

The reverse arguments apply as prices go above $20: at such prices the expansion of offshore exploration and production, and enhanced recovery of existing reservoirs, would continue, as would substitution by natural gas. Coal in the electricity market could bear some additional carbon or environmental costs and still compete. The long-term 'fundamental' band has probably not changed. Any change is likely to be downward rather than upwards as the infrastructure for offshore petroleum production and international gas transportation grows (see Chapter 5) and its capital costs are sunk.

Competition within the price band

The oil market is approaching a key shift between episodes, with the end of the large structural surplus of crude production capacity which has dominated the market, and all attempts to restrict competition, since 1973 (Figure 6.7).

The point is not that there is some mechanical connection between spare capacity and price, or that spare capacity is about to disappear. There is still a large overhang of reserves in several exporting countries; but the use of *short-term* market power to defend 'survival' levels becomes more credible if the spare capacity either can be absorbed by one or a few exporters, or is likely to be overtaken by demand within a short period of time. With the exception of Iraq, countries in the 'core' will not gain from higher volumes when prices fall. This market power does not work so well to sustain or increase prices: other governments and the private sector can increase oil production in the medium term. In the long term, competition from without would bring oil prices back into the price band.

The true lesson of 1998–2000 may be that when the capacity gap is narrow, correction needs to be quick as the market can slide between

Figure 6.7: Spare crude oil production capacity

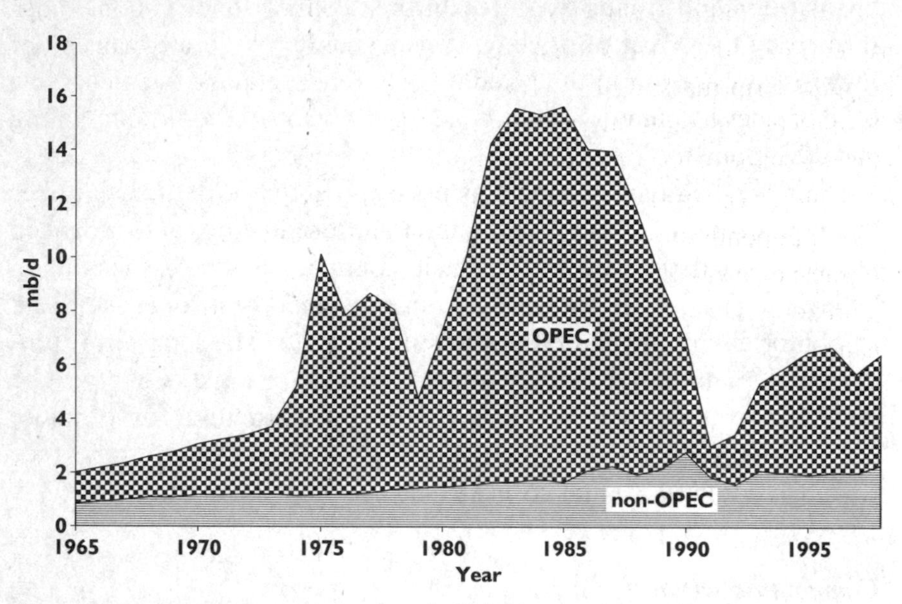

Source: IEA estimate; communication from A. Bieleski.

surplus and shortage and back before the slow negotiation of OPEC policy can change the degree of competition. In such a situation the Saudi Arabian government could play the role of a short-term supply manager, expanding its capacity ahead of the others, and increasing production when prices rise 'too high'. The Saudi capability to play a dominant 'market leader' role may be much simpler in future. The task itself should be easier, as the overhang of surplus capacity which destabilized the market during the 1980s and 1990s has largely disappeared. The conventional visions of oil demand and supply all require capacity to be expanded soon for supplies to grow to match demand. The key to market leadership is the ability to vary greater capacity more quickly and more cheaply than competitors. The larger the Saudi share of production, and the lower the overall capacity surplus, the less risk this carries of the type experienced by Saudi Arabia as 'swing producer' in the 1980s. Whether the Saudi government will recognize this change and seize this role depends in part on its being legitimized in some way through OPEC. It will also be necessary for the Saudi government soon to start influencing expectations by announcing plans to increase oil production capacity in such a way that the news does not destabilize the short-term market.

If market leadership does not emerge, the likelihood is that the new 'episode' will be one of short, sharp shocks in both directions. The elastic of the price band is probably pulled tighter. Whether it has moved upward depends on what happens outside the oil industry: in gas, in the transport sector and in the policy areas in which fossil fuels appear as threats rather than prizes. The implications of this are discussed in Chapter 9.

Chapter 7

Energy security

After the 1973 oil crisis, energy security was a thriving topic. It seemed an imperative in its own right – a chapter in the great book of security issues, a long way below nuclear weapons but certainly alongside strategic materials and technologies. One did not discuss autopart security. A central reason for this was that the 1973 oil crisis represented a triple threat:

- Day-to-day life was disrupted: there were lines of motorists in the United States waiting impatiently to refill their automobile tanks with gasoline.
- There was an economic threat: negotiations between OPEC and the transnational oil companies ceased and OPEC began increasing prices, apparently at will, against a prospect of the world running out of oil.
- There was a political threat: the Arab oil embargo of supplies to the United States and the Netherlands (the entrepôt for Europe), and month-by-month reduction of all exports, were aimed at securing a change in US and European support for Israel in the face of Egyptian and Syrian attacks as these countries tried to reverse their losses in the 1967 Arab–Israeli war.

Nearly 30 years later, energy security has moved off the title page and is at best a footnote to today's and tomorrow's global security issues. Energy ministries have been absorbed into trade or economics ministries. 'Energy policy' is an empty phrase in a global economy in which governments in developed countries are withdrawing from controlling or owning energy as well as other industries. Competition is commoditizing the fuels which compete with oil as it has long since commoditized oil itself. In the past 27 years the world has not run out of oil. Oil prices have not been sustained for long above their 100-year average – below $20/bbl in today's dollars. Since the end of the Arab oil embargo and production cuts early in 1974, there have been no more political sanctions by oil exporters. The trend is the reverse: no longer is the United States the object of oil sanctions; its Senate, administration or private investors variously lead sanctions against oil-exporting countries and companies which do business in them. The relevant geopolitics have been defined by the Gulf War of 1990–1 as dramatically as they were previously defined by the 1973 oil crisis. As a result, the energy security policies of the earlier period are as marginal to the present conflicts as the Maginot line was to the mobile warfare of 1940.

This chapter attempts a cool look at today's realities, focusing on the differences between now and the 1970s in the oil sector, the world economy and the geopolitical situation; and on the need to think more carefully about the objectives of energy security, in particular, to distinguish between the energy aspects of economic security and any energy dimensions to the security of national integrity and domestic policies, and foreign policies. It suggests that the idea that energy security can be improved by reducing import dependence is an obstacle to clear thinking. The growing 'dependence' of the United States on energy imports since 1970 has not prevented it from becoming the world's most powerful nation. Energy self-sufficiency and surplus did not prevent the collapse of the Soviet Union. The chapter concludes by suggesting strands in a more complicated network of security interests:

- the management of supply disruptions (an old subject) in the context of the growing Asian share of Middle East oil imports;
- reinforcing the physical infrastructure and market conditions for substitutability and competition among fuels;
- reviewing distortions imposed by taxes and subsidies which lead to inefficiencies in the use of fuel by discriminating between fuels and between sources of supply; and
- the potential for threats to trade which arise from essentially political conflicts to be reduced by attention to the political cause of the conflict, whether it is peace in Palestine or human rights in certain countries.

Figure 7.1: Sources of oil production, 1973 and 1999

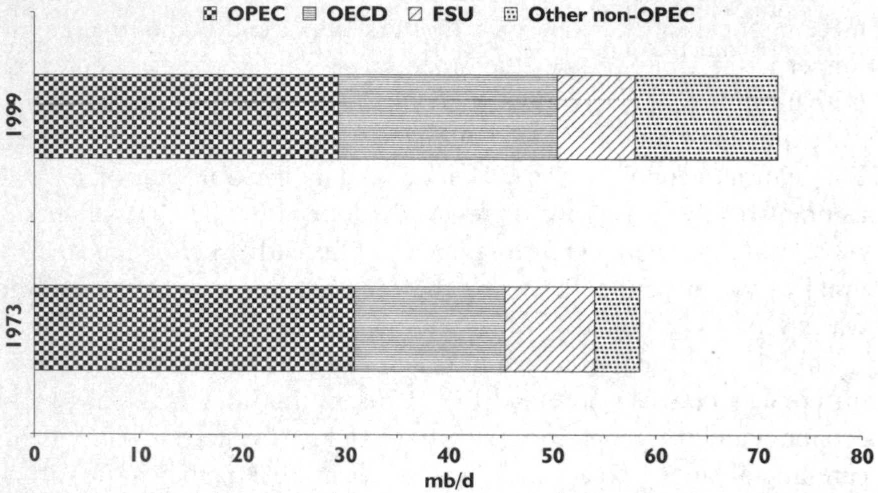

Source: BP Statistical Review 2000.

Oil now and in 1973

Diversification of oil supplies

Between 1965 and the first oil shock in 1973, the demand for oil worldwide grew at 8% a year. Oil supplied 46% of world energy in 1973 and its share was growing rapidly. So was the OPEC share of oil supply, which reached over 50% in 1973. For the 10 years to 1999, oil demand grew at just 1% annually. The oil share of the energy market worldwide has shrunk – it was down to 41% in 1999. The OPEC share of oil production, though it has risen since 1986, was in 1999 still only just over 40%, and the Middle East accounts for just over 30% of world oil consumption (13% of world energy consumption). Energy supplies have been diversified by the development of natural gas and nuclear energy, while oil supplies themselves have been diversified by discoveries and new developments in a wide variety of countries and continents. Only towards the end of the 1990s did OPEC regain its 1973 level of production – in an expanded market, as Figure 7.1 shows.

The conventional vision described in Chapter 2 sees some increase in the future. The EIA (in *International Energy Outlook 2000*) is the most optimistic of these observers in respect of the expansion of conventional oil, and particularly conventional oil in OPEC. In its reference projection, OPEC would have 51% of world oil production capacity in 2020, and the Middle East 41%. The relevance of these proportions to security in the current era is discussed below.

Structure of the world oil market

We are accustomed now to a global oil market in which prices are transparent and trade is free of quantitative restrictions and heavy tariffs. Oil is traded like any other commodity. There are institutional spot and futures exchanges in London and New York for crude oil and oil products. These prices determine the price of most imports and exports, even if volumes are set by long-term contracts. Most crude in international trade is traded between producers and traders or consumers who are independent of one another. Within the major importing regions there are no quotas or price controls. Russian producers compete to export oil. Prices of oil on the domestic market are moving towards international levels in Russia and China, and some other developing countries. Natural gas has gone the same way in North America and is beginning to follow suit in Europe. The effect is a connected set of commodity markets where competition is the rule and economics work. Supplies move from where they are cheapest to where consumers are prepared to pay the most for them. Investments follow the markets.

The transition to a competitive international market for oil took place between 1973 and the mid-1980s. It has left behind the state oil producers of OPEC. OPEC's attempts to manage the market by collective action have been destabilizing: increasing supply in the face of falling demand in November 1997, and failing to match rising demand in the first half of 2000. In March 2000, OPEC attempted to provide a quicker response so that production would automatically adjust when the price of the 'basket' of OPEC crude fell outside the range of $22–24 per barrel. This mechanism broke down on its first test, in July 2000. The idea of concerted supply intervention when prices move outside a prescribed band is not ridiculous: it was the basis of some UN commodity agreements (such as those for wheat and sugar) which can be compatible with GATT. However, such agreements require wide coverage of producers and consumers, and for the negotiation of price brackets between governments representing major importers as well as exporters. For OPEC governments, this would involve turning the clock back to before their assumption of control over their price and production policy in 1973. For the importing countries, such agreements would take them back into the economic ideologies of state intervention of 20 years ago.

Before 1973, the situation was different at every point. Something like 90% of international oil trade was conducted by a handful of companies – the 'seven sisters' – which had almost exclusive rights to lift oil from what were then the main exporting countries. These companies also dominated the refining and marketing sectors of Europe and North America. Prices for crude oil in international trade were based on long-term tax reference prices which were set by agreements between the companies and the exporting governments. These agreements were finally destroyed in October 1973 by OPEC's decision to set prices unilaterally in the future.[1] Prices of oil products in the main importing countries were already under government control as a result of anti-inflationary policies such as those imposed by the United States in 1971 to deal with general economic problems. Natural gas was also under price control in both the United States and Europe, where it was only just beginning to appear as a major source of energy. Coal was, in many countries, under the effective control of miners' unions, as the UK found to its cost in 1973. The 'three-day week' and the power shortages which brought down the government of Mr Edward Heath in 1974 were caused by dependence on British coal, not by dependence on imported oil.

[1] For a step-by-step account of the interaction between supply cuts and pricing decisions, see Gavin Brown, *OPEC and the World Energy Market*, 2nd edn (London, Longman, 1990), pp. 439-45.

The changing role of governments in the global economy

The developments in the energy industry between 1973 and the end of the century took place in the context of wider economic changes which went in the same direction. Governments removed price controls across the OECD economies generally. Regulations were redesigned to promote competition, rather than to control it, in most industries. Many governments privatized producing assets and stepped out from the micro-management of the economy. Nor was this all. In North America and Europe the so-called Reagan–Thatcher years saw many governments withdrawing from Keynesian-type macroeconomic policies. Fiscal discipline became the order of the day. For the members of the European monetary union this was entrenched by treaty into the conditions for participation in the single European currency. Non-OECD countries have gone the same way. Some of them, like Mexico and Korea, have joined the OECD. Others, like Argentina and now Brazil, have followed the new orthodoxy, supported, more or less, by the international financial institutions. National deregulation has gone hand in hand with the development of further global agreements about non-intervention in the markets for goods (the Uruguay round of GATT), services (General Agreement on Trade in Services or GATS) and finance (Trade Related Investment Measures or TRIMS).

The effect of these macroeconomic changes has been, for many oil-importing countries, to remove the framework within which it was possible to give a national 'energy policy' the kind of meaning it had in the 1970s. Bilateral agreements between countries about energy supply, where they existed, have disappeared. Government investment in energy projects provides export credit or political insurance for the companies involved and does not implement national energy policies for the investors. (The acceptability of these projects may be affected by national values on issues such as environmental practice and human rights, as discussed in the next chapter.) In today's world, reading old energy policy statements is like reading the medieval mystery plays: the language is archaic; the plots centre on the same struggle between good and evil, but, unlike the medieval audiences, we no longer recognize our neighbours in the characters.

Figure 7.2: Current account oil revenues as a percentage of non-oil imports, 1980 and 1997

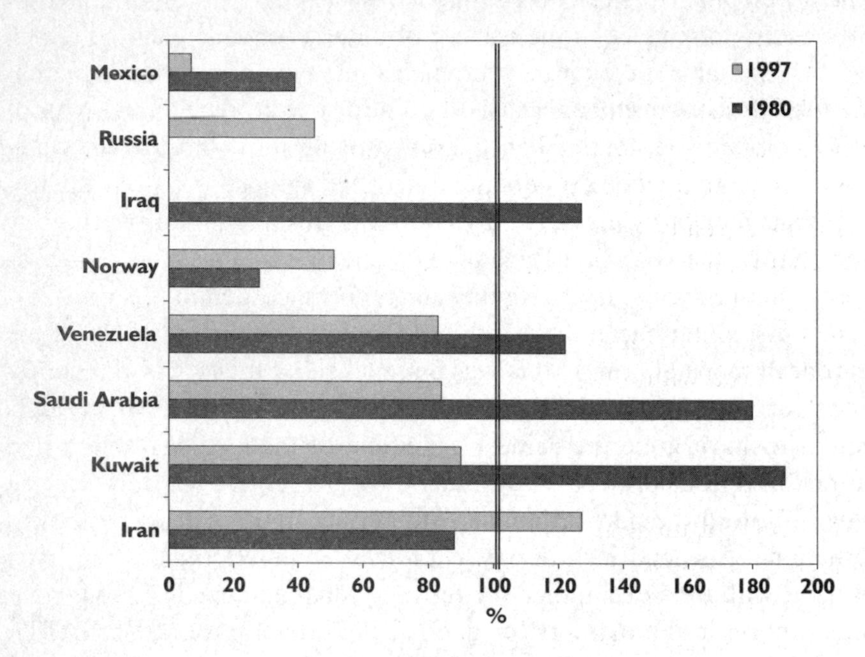

Source: World Bank, *World Development Indicators 1999*, Tables 4.5, 4.6.

Governments are now promoting competition, not planning investments in energy. Budget deficits are bad politics, not good Keynesian theory. Energy security has to be placed within the context of a global economy which not only is interdependent but in which political and legal institutions either accommodate or reinforce that interdependence.

Similar trends are taking hold, of necessity, in the oil-exporting countries. Countries like Saudi Arabia were traditionally the 'low absorbers' of revenue, capable of withholding oil production and losing market share in order to support the price of oil. All are 'high absorbers' now, as Figure 7.2 shows. In 1980 Iraq, Venezuela, Saudi Arabia and Kuwait all generated current account surpluses. By 1997 (the last year before the price fluctuations of 1998–9) only Iran was generating surpluses – because of internal and external constraints on economic growth and the necessity to service debt. For the main OPEC exporting countries, import capacity depends on oil export revenues.

The geopolitical context

While the global economic context for energy has changed since the 1970s, so has the global security environment. In most countries energy has already been downgraded from a chapter to a footnote in national security policy, but security policy itself is back with the editors. Any new writing affecting energy security needs to take account of the collapse of the Soviet Union, changes in the Middle East, and the rise of China within Asia and as part of the world geopolitical system.

The Gulf and the superpowers

The first and most obvious geopolitical phenomenon of the 1990s was the collapse of the Soviet Union. It explains, among other things, why US intervention in the Middle East was so risky and limited in 1973 and so overwhelming in 1990. In 1973, the Soviet Union was a leading supplier of arms to Egypt and Syria, the two countries initiating the attack on Israel at that time. When, with US aid, the Israeli army halted the Arab advance, the threatened Egyptian army was supplied by the Soviet military. Soviet threats led to a worldwide US nuclear alert. Further military escalation was prevented by a rapid ceasefire agreed under pressure from both the United States and the Soviet Union, mediated by Henry Kissinger. The Arab embargo on exports of oil to the United States and the Netherlands, and the production cuts that reinforced it, continued with no US military intervention, until March 1974.[2]

By 1990, the situation was already different. Perestroika began in 1987. The Berlin Wall came down in 1989. In June 1990 the Russian parliament asserted its sovereignty against the government of the Soviet

[2] For a vivid account of the connections between the war, the breaking up of the company–government oil price negotiations, and the embargo, see Daniel Yergin, *The Prize* (New York: Simon & Schuster, 1991).

Union, which ceased to exist in December 1991. Saddam Hussein's errant sense of timing led him to attack Kuwait and threaten US global interests just at the moment when the Soviet Union was collapsing. It is debatable whether US-led UN intervention in Kuwait would have occurred if the Soviet Union had still had the capacity to confront the United States on the question of how to respond to the Iraqi aggression, rather than having to come to an accommodation. Nevertheless, through its membership of the Security Council and the leverage of its vote there, the Soviet government was able to exercise some influence on the terms of the UN intervention and sanctions on Iraq after the latter's attack on Kuwait in October 1990. The circumstances of 1990 were exceptionally favourable to the idea of UN intervention. Saddam Hussein appeared to threaten all Gulf countries: they had no alternative source of security. The apparent direct threat to oil supplies justified European and Japanese support. There was no direct Chinese interest. The Soviet government had supplied arms to Iraq before the war. As sanctions mounted, it was slow to withdraw its technical support and was never paid for its supplies.

Ten years later, much has changed in the countries of the FSU, but the Russian Federation is in a similar weakened position relative to the Middle East. It has a Security Council vote; it can supply, or threaten to supply, some military technology, in particular nuclear technology; but it cannot afford either to supply goods and services on credit or to mount any military action of its own in the area. Russian privatized oil companies have been leaders in signing memoranda of understanding for investment in Iraqi oilfields but very little appears to have happened: they too lack the financial resources and the technology that Iraq needs most for its ambitious oil expansion programme.

The diplomatic and military success of US/UN intervention in 1990 profoundly changed the security position in the Gulf from that which had prevailed in the 1970s and 1980s. It is the icon for a new era in which the United States has come to be perceived as a dominant power in the region, capable of decisive military action there to protect its interests. This is a complete contrast to the dismal humiliation of the United States in 1979 during the Iranian hostage crisis.

Caspian oil and gas

Much has been made by some journalists of the so-called 'great game' of intrigue among Russia and foreign powers for dominance of central Asia, with oil now the centre of attention. This exaggerates the importance of Caspian oil. Production is currently about 0.75m b/d, about 1% of world production. Exports are perhaps one-third of that. The most ambitious figures (suggested by the US EIA in 1998[3]) approached 7m b/d by 2020; the EIA's latest estimates are more conservative about the speed of development, suggesting 2m b/d by 2005.[4] The discovery of a probably significant oilfield at Kashagan, offshore Kazakhstan, keeps alive the hope of significant new discoveries in the North Caspian. It remains arguable whether enough reserves will be proved soon enough to support such production and therefore to justify building a major pipeline to export 1m b/d through Turkey – as the US government advocates. The fact is that optimistic guesses about oil resources in the Caspian have not yet been verified, and there has not been much drilling. Gas discoveries have been better, but not necessarily more rewarding. Gas exports from Azerbaijan need new pipelines. If gas exports from Turkmenistan through the existing Russian pipeline system are to be revived, new markets are needed.

Russian experts have always expressed a poor opinion of the hydrocarbon potential of the Caspian,[5] and it is unlikely that Russian policy towards these countries is driven by hopes of recapturing great oil wealth. There will be tariff revenue to be gained from these pipelines and this will be valuable to transit countries like Georgia. Pipeline fees are not likely to be a prize the Russian Federation should find worth fighting for. Russian interests in the region are both broader than oil and simpler. At the minimum, Russia has an interest in preventing these newly independent countries from falling under the domination of any other regional power, for example, Turkey or Iran, or becoming a new frontier for the so-called hegemony of the United States. At the

---

[3] EIA, *International Energy Outlook 1998*, p. 34.

[4] Ibid.

[5] A. Konoplianik, *Caspian Oil at Eurasian Crossroads: Preliminary Study of Economic Prospects* (Moscow: privately published, 1998).

maximum, Russia would seek dominant influence over these countries' domestic as well as foreign policies. Rapid development of Caspian hydrocarbon resources, even though they may be modest in global terms, is not in Russia's interest because even modest oil and pipeline revenues will strengthen the independence of these small countries and make them worth the sponsorship and patronage of the United States and regional powers.

The legal status of the Caspian Sea is still unresolved in many respects. This uncertainty will increase the cost and reduce the availability of foreign investment, for example by banks and multilateral lending agencies. Russia can offer alternative export routes, often through existing pipelines or by relatively inexpensive expansions. These alternatives challenge the economics of the multiple routes canvassed by the United States to avoid Russia.

The limited nature of these opportunities suggests that the geopolitical 'game' in the Transcaucasus and central Asia is not about securing energy resources either for importing countries or for Russia; the prize is not worth the likely costs. It is about the degree of independence that the taxes and royalties from oil and gas resources may bring these countries in the future. Energy is a means, not an end. Russia also has many cards to play short of military action.

As a practical matter, external military intervention in the Transcaucasian and Caspian countries is really an option for only one country – Russia. Non-military factors deployed by the United States and other countries could make that a very costly option. The 'best' that can be hoped for, in terms of independence for these countries, is a replica of the position in the Gulf before 1989, where US intervention was limited – as in the case of the flagging of Kuwait tankers – to very specific action responding to very specific local threats to the interests of the United States and its allies. For the Caspian, the analogy would be the protection of export pipelines within the independent states against threats which were generated for local reasons and not explicitly promoted by Russia.

One sad consequence of this scenario would be that the Caspian and Transcaucasus countries would have to maintain high defence budgets, like the Gulf countries, to provide security both against one another, and against proxy wars and subversion. This would minimize their dependence on an uncertain 'balance of non-intervention' from outside. This may be good news from the point of view of encouraging those countries to support rapid expansion of oil and gas exports to buy arms; but it is not necessarily good news for the development of prosperous and stable democracies in the area.

Russian gas

There are less prominent but similar issues for Russia as an exporter of gas, as discussed in Chapter 5. Exports of Russian natural gas are an important generator of hard currency for the Russian government. They are completely dependent on the marketing of that gas in western Europe, where it supplies just under 30% of natural gas consumption and about 6% of total energy consumption. Looking from the Russian end of the pipelines, this west European export market accounts for about 25% of Russian gas production. For Russia, these are pipeline exports for which there are no alternative markets in the short and medium term.[6] Through Gazprom, Russia is planning to diversify its gas export routes to Europe by constructing new pipelines through Belarus and Finland, to minimize dependence on transit through the Ukraine. For western Europe, the developing gas grid, with import terminals for Algerian gas and for LNG, provide some short-term and also larger long-term alternatives. In the long run, Russia needs the European gas market and the European gas market does not need to pay extortionate prices for Russian gas.

The Middle East's own agendas

Since the 1970s, the geopolitical situation in the Middle East has also changed substantially, though not entirely. The stability of relations

[6] In the long term some limited market might develop in China, but the medium-term prospect is for Chinese gas imports to come from currently undeveloped fields in east Siberia and Central Asia, not from the fields in north Siberia which are linked to western Europe.

among the Gulf countries and within some of them still cannot be taken for granted. But, on the positive side, there is an intermittent peace process for Palestine, and there is a peace treaty between Egypt and Israel. The position in Iran is quite different from what it was 20 years ago. The pluralist system of government there continues to evolve: elections produce changes; checks and balances work. On the negative side, Saddam Hussein is still in power in Iraq, which is regularly bombed by US and UK forces under UN auspices. The nature of any successor regime is problematic. The capacity for peaceful change does not appear to have been institutionalized in other Gulf countries except by reliance on the flexibility of their monarchies. Can we put these developments in a broader political context in looking at the future of energy supplies from the area?

Two elements can be identified. The first is the existence of the peace process (not present in 1973) and its continuing momentum. The Arab oil embargo of 1973, short as it was, was a response to an Arab–Israeli war. Such a war would probably still be the only cause which could unite the Middle Eastern oil exporters to attempt to repeat an embargo. Today, it is less likely to come about. Moreover, the effect of such an embargo would be less damaging and less frightening to the oil importers than in 1973 because the structure of the oil market is now more flexible and diversified.

The second element is less positive. These countries have the most rapid rates of population increase in the world, and their demands for revenue are growing even faster. The accumulated surpluses of the 1970s and early 1980s have been spent – sadly, much of them on armaments and war: first the Iran–Iraq War and then the Gulf War, which was essentially paid for out of the accumulated financial surpluses of the Gulf producers.[7] This history of war in the Middle East, and the countries' evident preference for maintaining so-called defence expenditure, is vivid evidence that the oil exporters of the Middle East are not just commercial competitors in the oil market. They are geopolitical rivals. Their rivalry is based not only on economic interests and the

[7] Whether Iraq will ever pay the contribution demanded from it by way of reparations is another matter.

struggle for power in the region but on differences among them in respect of values and religion.

Stability in the Gulf might perhaps be imposed again from outside if it were upset from inside. But the experience of the Gulf War may be giving a misleading impression of the extent to which external military intervention can protect energy supplies in the future. A less generally offensive threat than Saddam may appear, so that international support may be difficult to achieve. Internal disruptions – as in Iran in 1979 – may not justify intervention. The possibility of overwhelming victory may be much more remote. The Kosovo model of intervention without ground warfare has set a new standard of intervention with minimum risk to foreign personnel which may have correspondingly limited effectiveness.

Insecure exporters

It is worth here making the obverse point about the dependence of the oil exporters on their exports. Unlike the other major oil-producing countries – the United States and Russia – the major Middle East producers are dependent on export markets for most of their foreign currency earnings, a high proportion of their government revenues, and through that, for a large fraction of the national income. (The situation varies from country to country, and good figures are hard to come by. Middle Eastern countries as a whole export over 70% of their oil and gas production.)

Instability in the international oil markets creates real economic insecurity for the Middle East exporting countries. With real economic insecurity comes real political insecurity: can these governments meet the steadily growing demands of their rapidly growing populations? Can they match the efforts of the rival neighbours to achieve regional influence, or to promote competing religious outlooks, or different approaches towards secularism, towards the organization of the state, towards foreign alliances and foreign investment? These countries are very different, not only in their geopolitical interests and domestic values, but in their oil potential. In simple terms, some have more barrels per head than others. Figure 7.3 compares their oil reserves with their

Figure 7.3: Oil reserves versus population, 1997–98

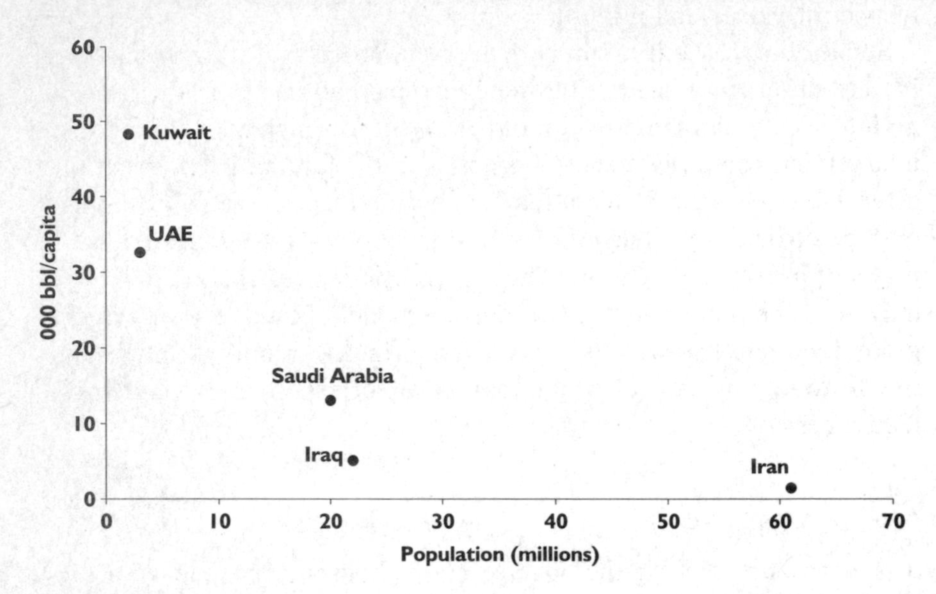

Source: Oil reserves data are taken from BP Amoco: *Statistical Review 1999*. P 4. Population data from World Bank: *Development Indicators 1999* Table 1.1.

populations. Countries with high reserves and small populations may be more relaxed about price because of their access to volume. This means that some countries, like Saudi Arabia, can grow their revenues even under low prices by increasing production; others, such as Iran and Abu Dhabi, do not have that physical capacity. This simple fact is what makes it so difficult for the Middle Eastern oil exporters to agree on output and pricing policies except under the most extreme conditions – such as an oil price of $10/bbl in 1998.

From economics to security for importers

In Chapter 6 we argued that the differences between current capacity and potential, combined with differences in population and in geopolitical rivalry, were such that Middle Eastern exporters are bound to compete with one another except when low oil prices threaten the stability of the governments of all the principal exporting countries.

In the medium term, competition among exporters, and in the longer term, competition from natural gas and other fuels would prevent cartelization of oil prices. This is not to say that oil prices will not rise, only that any rise would be for economic reasons rather than as a result of intervention by the Middle East exporting governments. There might be economic issues for the oil importers, but not a security issue.

These conclusions do not eliminate economic risks to imports: there can be unintended disruptions in the future as there have been in the past. Investment in expanding oil production capacity may slow because of shared perceptions about the merits of postponing it. Investments in competing fuels may be slowed by failure to overcome the obstacles to developing new markets – for example, natural gas in China and India – which depends on the development of a physical import infrastructure. These are economic risks which apply to most commodities and which go with benefits of international trade. Domestic supplies might carry a different set of risks, as well as higher costs: exposure to domestic market power – such as trade unions and property owners; loss of competition and flexibility; and lack of diversity.

Economics matter for national security because the strength of the economy affects the ability of a country to sustain a military defence policy and to support its foreign policy with economic and, if necessary, military means. But from the security point of view the important question is not whether the risks of energy trade with the Middle East can be avoided, but whether importing countries which are industrial or political rivals will be differently affected, and how big the effects might be in relation to the alternatives. Figure 7.4 gives a broad picture of the importance of Middle East oil in terms of world energy and oil supplies, and in the international oil trade now and in the 'conventional vision' for 2020 – taking the EIA reference (*International Energy Outlook 2000*) view of Middle East oil production capacity.

In absolute numbers, the Middle East supplies about 26% of the world's liquid fuel supplies and about 10% of the world's energy supplies. By 2020, using rather positive assumptions about growth in oil demand, these percentages could grow to about 36% of oil supplies

Figure 7.4: Middle East shares of markets

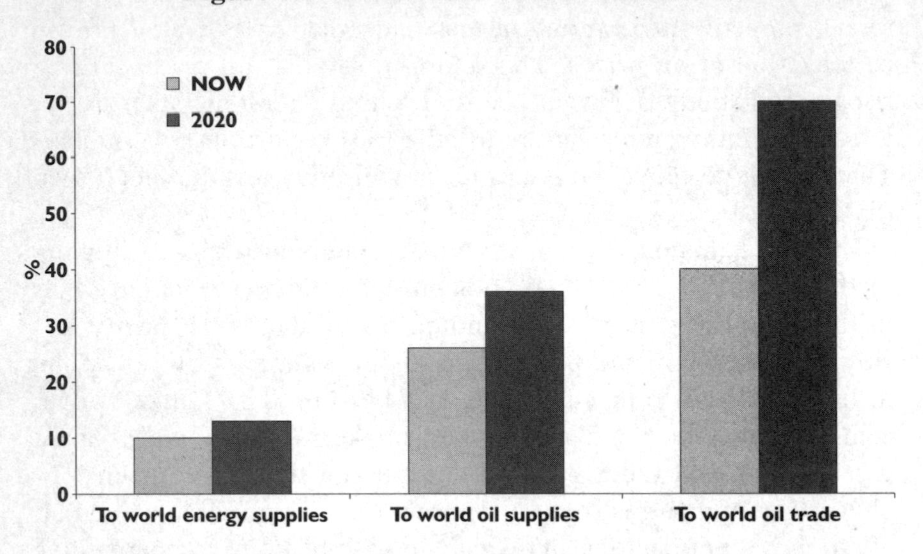

and about 13% of world energy supplies.[8] Today the Middle East oil accounts for about 5% of US energy consumption and about 10% of European energy consumption. For 2020 the proportions will be roughly the same. In the long run, this level of dependence should not be regarded as life-threatening to the world economy or as carrying intolerable economic risks to energy users.

For inter-regional oil trade, the Middle East is more important: it supplies about 40% of the world's inter-regional oil trade today and on the same projections would supply about 70% in 2020. However, there is a shift from west to east in imports of Middle East oil. The East Asia and Pacific region already takes over half of Middle East exports, and this fraction will rise to over 60% by 2020 with the growing Asian share of world oil imports. The interdependent oil trade between West and East Asia is becoming a new axis for security concerns.

For the major developed countries the 'exposure' to the unavoidable economic risks of the oil business falls on economies which

[8] These percentages are based on the reference case of the EIA in *International Energy Outlook 1999*. The projections in the IEA's *World Energy Outlook 1998* suggest a slightly lower Middle East share of world oil trade.

Figure 7.5: The value of net fuel imports as a percentage of non-fuel exports of goods and services

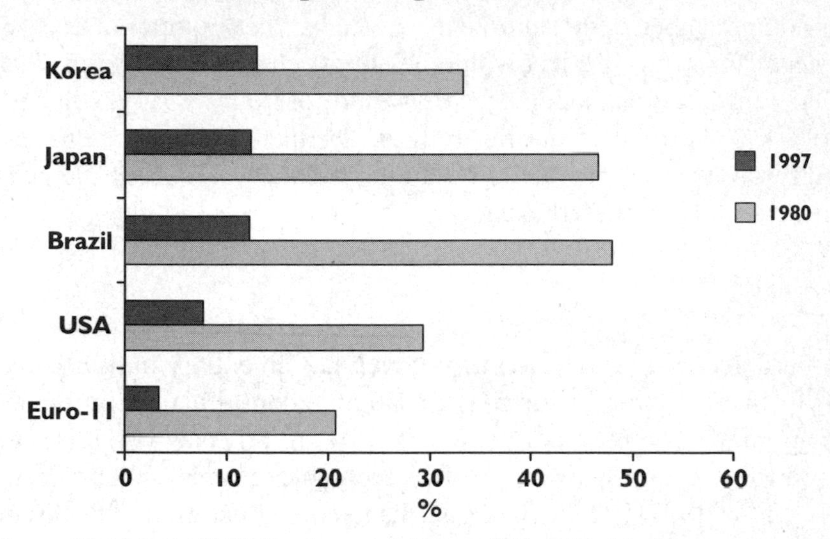

Source: World Bank, *World Development Indicators 1999*, tables 4.5, 4.6.

have adapted to the risks of trade and have different capacities to carry them. Figure 7.5 is the reverse of Figure 7.2. It shows, for importers, the fraction of non-energy exports which balanced the cost of energy imports in 1980 and in 1997, the last year before the 1998–9 swings (with an oil price around $20 (1999$)/bbl). This is a snapshot (before reactions to price changes set in) of the static effect on the terms of trade – and therefore on the GNP – of the countries concerned. It reflects (quite properly) the value and volume of fuel imports relative to the value of other trade.

There are two striking things about this chart. For Japan, Korea and Brazil, countries in different stages of development, the 'burden' of fuel imports was in 1997 very similar (with the Japanese economy out of recession, its number would probably be lower). But compared to them, the United States dedicated less of its export capacity to fuel imports, and for the euro-zone the number is even lower.[9] The impact of

[9] Broadly speaking, with equivalent imports per unit of GDP, the terms of trade effect will be less for countries where trade is higher relative to GDP: the impact will be spread over a larger export capacity.

a shock change in price would fall hardest on the terms of trade of the first three industrial competitors, and least on the euro-zone. For the UK, not a member of the euro-zone and a small net exporter of energy, the net terms of trade effect would be almost entirely relative: the UK would gain because its competitors would lose during surges in fuel prices, and the reverse during troughs. Of all the leading industrial country group, Europe should be the least concerned about the economic risks of oil price fluctuations.

Asia again

The third geopolitical development which has an energy impact is the growth of the Asian economies and their geopolitical importance. Among many complex developments, two are especially significant: the growth of China, helped by market economic reforms; and the financial crisis of 1997–98. The latter has had some effect in binding Asian countries more closely into the developing global market-orientated forms of economic organization and in weakening the role of the governments, state industries and government client firms in the economy. Underlying the growing importance of Asian countries in the world economy and energy markets is the continuing existence of rivalries and conflicts within the region, and the relative absence of institutions for adopting cooperative approaches to security issues – including specifically the managing of supply disruptions.

Asian countries together already consume about 30% of the world's energy, as much as North America and 50% more than Europe.[10] The same factors apply to oil. In 1998 the Asia–Pacific countries consumed more oil than the United States or Europe. Even after the 1997–8 crisis, projections for 2020 suggest Asia will then consume around 40% of the world's oil and 40% of its energy. Over half of Middle Eastern oil production and 60% of its exports already goes east, rather than west. These proportions will rise as Asia takes a larger share – indeed, the greater part of world oil trade in the future.

[10] Including the Middle East, which accounts for 6% of world oil and 4% of world energy consumption.

In the panorama of Asian scenarios, three cast their shadow over global energy. The first shadow rolling across divided Asia is the growth of China as an energy importer. It has an interest in avoiding both dependence on Middle Eastern oil and import routes open to domination by the US navy and airforce. The Chinese government and state companies are showing a strategic interest in the import of natural gas and oil by pipeline from eastern Siberia and Central Asia. Gas pipeline developments could significantly reduce the growth of imports of Middle Eastern oil into China. The effect on the world oil supply–demand balance could be at least as great as the more realistic figures projected for Caspian oil.[11] However, the commitment to the infrastructure investments necessarily depends on a degree of political support from the two governments concerned, sufficient to attract the foreign investment necessary for the development of the Russian gas reserves and probably the financing of part of the infrastructure and market development costs. Put simply, are the two governments and potential investors confident that future tensions between Russia and China would not affect the building and use of the pipelines? European countries took that kind of decision about building pipelines to import Russian gas in the early 1980s. The bet was that the benefits to both sides of using the pipeline would be such that it would not be interrupted for Cold War purposes, while the probability of 'hot war' was so low as to be discounted.

The second shadow – long-standing but growing in length – is the absence of agreed mechanisms to deal with disruption of supply. Japan, Australia and New Zealand are the only Asian countries that are full participants in the IEA emergency sharing and stockpiling schemes.[12] Moreover, among the major importers, only Japan has fully liberalized its oil import and pricing system. A disruption of oil supplies to East Asia would be accompanied by acute local shortages in countries highly dependent on oil imports (Korea imports all its oil,

---

[11] See John V. Mitchell and Christiaan Vrolijk, *Closing Asia's Energy Gaps*, Briefing Paper (London: Royal Institute of International Affairs, 1998).

[12] Sharing mechanisms can be triggered if a single member country experiences a 7% 'shortfall' in its normal supplies of oil.

Thailand over 90%).[13] The pricing mechanisms to deal with such disruptions would be likely to be hindered by government attempts to shelter consumers from the effects of scarcity pricing.[14]

The Asian countries lack an institutional structure within which to address the question of energy disruptions, or indeed energy security generally. The Asia–Pacific Economic Cooperation forum (APEC) and Asia–Pacific Energy Research Centre (APERC) address conventional energy issues, not the geopolitical context or the growing issues of environmental, social and political impact. These are an element in the tension in Indonesia between the central government and the islands, where oil and gas production, and export facilities are located. Indonesia supplied LNG for 36% of Japanese and 64% of Korean gas consumption in 1999. For the Asian importing countries, the idea of developing favourable bilateral relations with oil-exporting governments seems to provide few opportunities. There are real difficulties in creating advantageous bilateral relationships in a global and transparent commodity market. The practical opportunities for reciprocal investments seem limited. In 2000, the Saudi Arabian terms for renewing the Arabian Oil Company (a Japanese company) concession in the neutral zone were such that the company withdrew and the concession reverted to Saudi Aramco. There has been Japanese government support for company interest in new openings in Iran upstream.[15] Japanese companies have a significant, though not large, presence in Indonesian oil development and production and are partners in LNG export projects in the Middle East, Indonesia and Australia. The Chinese National Petroleum Company bid successfully for participation in Venezuelan projects in 1998, and has negotiated production-sharing contracts for oil and gas in Kazakhstan and Turkmenistan. Saudi Aramco has invested in a Korean refining company, but has been rejected as an investor in a Chinese refining project and has failed to achieve a commercial investment in

[13] See Mitchell and Vrolick, *Closing Asia's Energy Gaps*.

[14] For a detailed analysis of Asian exposure to disruptions of oil supply, see Ken Koyama, 'Oil Supply Security in Asian Economies', *Energy in Japan*, no. 149 (Japanese Institute of Energy Economics, 1998) pp. 31–47.

[15] 'Tokyo Sets its Sights on Iranian Upstream', *Petroleum Intelligence Weekly*, 1 May 2000, p. 3.

the (generally unprofitable) Japanese refining industry – although foreign investments in refining are now free of restriction. The flow of upstream and downstream investments will continue as investment conditions ease, but it is difficult to see these making a material change in the exposure of Japan to the risks of supply disruption. Their significance is more in the increasing exposure of Asian energy companies to the rigours of international competition, direct dealings with exporting governments, and the scrutiny of international public opinion.

The third shadow is cast by the relations between Asian oil and gas companies and their home country governments and public opinion. Accountability here is different from that of US and European companies. At the simplest level, regional companies may not be as concerned about environmental and human rights issues as companies accountable to the public in Europe and North America.[16] But accountability to local government and public opinion is only part of corporate responsibility. Chapter 8 describes how the dependence of even state-owned companies on access to US and European financial markets, and on the multilateral lending and credit agencies, exposes them to the growing concerns about the social and environmental impacts of major projects. There are a number of these controversial projects in Asia, in hydroelectricity, gas pipelines and oil developments.

Objectives of energy security

Summary of economic risks

In the new situation described above, the traditional objective of 'security of adequate supply at fair prices' is not meaningless, but it is not useful. It begs the question of what is adequate and what prices are fair.[17] It does not easily translate into a plan for action in the oil, economic and political circumstances of the beginning of the twenty-first

[16] The replacement of Texaco by Petronas as an investor in the Burma gas pipeline project is an example.

[17] St Thomas Aquinas defined a 'just price' as the price which would prevail in a market if there were a market (Palgrave's *Dictionary of Economics*), which does not help much either.

century; the gains from international trade are too great for both importers and exporters to make reducing import or export dependence a catch-all security policy.

The first key to a new framework for energy within a general security policy is to distinguish economic from political risks. The argument of this chapter so far has been that long-term economic risks – essentially the risks of underinvestment in supply and cartelization of exports – are limited. Oil producers will continue to compete with one another and with other fuel producers (and suppliers of energy-saving systems).

The real economic risk, which all importers share, but in different degree, is that some of the export supplies concentrated in the Middle East could be temporarily disrupted by new local or internal conflicts in that area. The IEA emergency stockpiling and sharing schemes provide a simple basis for policy response. (The IEA requires member countries to hold stocks to cover 90 days of oil consumption. If all countries did the same, this would arithmetically cover a year's total loss of all Middle East supplies.)

National security risks

At the most basic level, a nation might look for security to support its military and essential civilian needs to protect its independence in time of war. What constitutes such security depends on one's view of the enemy and of the likely nature and duration of the war. Nations seldom plan for perpetual war or even long wars, so that the kind of security required is short term. The argument is about how much, rather than about whether, military stock should be, what levels of compulsory storage of fuels are sufficient to match essential needs, and whether installations and transportation are adequately protected. In the case of security against the British coal miners it was important whether coal stocks were held at the power stations or at the mineheads. Although the problem is technically quite complex, the argument is conceptually quite simple: How much is the country prepared to pay for strategic reserves?

Where there is dependence on dedicated supply, as with a pipeline, one has to ask the political question of the probability of war between the countries at either end of the pipeline or along the route. It is easier to answer that question in Europe or the Americas than in Asia or Africa.

Securing domestic policies

At the next level, a nation might look to protect its freedom to manage its internal affairs. UN sanctions against oil supplies to South Africa during the apartheid era, and US or UN sanctions against Burma and Sudan in 2000, were intended to affect domestic policies in the sanctioned countries particularly with respect to human rights. Countries (or companies doing business in them) may also be sanctioned by investors or by consumers for human rights or environmental reasons. Energy trade, and dependence on energy exports, may be a target in such cases.

Apartheid was not brought down by energy sanctions – though other international sanctions, especially on lending, may have contributed to its demise.[18] South Africa achieved minimal energy import dependence in an economy endowed with sufficient cheap coal to power the electricity sector and provide an expensive but significant supply of synthetic liquid fuel. What oil imports remained necessary were obtained from minor exporters at premium prices. South Africa's reason for incurring these economic costs was not fear of an oil cartel but the difficulties it foresaw in defending political institutions which were universally condemned by governments and the public opinion of major countries outside South Africa and by the majority of members of the UN.

The lesson from South Africa's near 'energy independence' is that if a country expects its domestic policies to be permanently offensive to powerful countries, then it needs to prepare for long-term reductions in its sanctionable dealings with those powerful countries. It may look for other trading partners, or it may seek to reduce dependence on trade,

[18] See Jack Spence, 'South Africa: A Case Study in Human Rights and Sanctions', in John V. Mitchell, ed., *Companies in a World of Conflict* (London: Royal Institute of International Affairs/Earthscan, 1997).

or may attempt some combination of the two – as was indeed the South African position during the apartheid regime.

Such considerations may be an element in policy debates in some capitals today. It would be surprising were they to attract big budgets for energy independence in Washington or the capitals of other OECD countries.

Securing foreign policy

A further level of security concerns a country's freedom to practise an independent foreign policy, or at least to avoid having other countries impose constraints on its foreign policy by threats to deny it energy supplies or markets, or to make access to them damagingly expensive (or unprofitable).

Because of the position of the United States in global geopolitics, and in the politics of the Middle East, the first question is whether US foreign policy could be put at risk by any action on the part of oil-exporting countries. The US commitment to support the continued existence of Israel, and the continuing failure to conclude internal and external peace settlements for Israel, mean that this risk to US foreign policy has not been completely eliminated. The possibility of a repetition of the unsuccessful Arab oil embargo of 1973, which was designed to diminish international support for Israel, then at war with Egypt and Syria, cannot be excluded completely. But (as was the case in 1974) the capacity of oil exporters inherently hostile to Israel to sustain an oil export sanction is limited. To be effective the sanction would have to take the form of a general cutback in production, which the experience of 1974 shows cannot be sustained to the economic advantage of the exporters. For a political embargo to be effective, it would need to be big, and would as a consequence trigger the IEA mechanisms for stockpile use and emergency sharing. In the fluid oil and gas markets of today, the inevitable repercussions for oil demand and revenues would be more rapid. For the exporters such a confrontation would be hugely costly, to be undertaken only as a last resort in some cataclysmic crisis involving Israel in such a way as to unite the principal exporters. Finally, the geopolitical context is different. The United States would have more freedom to use other levers to avoid or mitigate the force of any such cutbacks.

For the United States at the turn of the century, political sanctions on oil and energy represent a foreign policy opportunity, not a threat. Support for US foreign policy is the prime goal of the US current energy-related embargoes against the so-called 'rogue states' of Libya, Iran, North Korea and Afghanistan, and of the UN sanctions against Iraq. The sanctions against domestic policies in Burma and Sudan express also a missionary aspect which is a continuing feature of US foreign policy.

How long term are these conflicts? There have been changes towards normality in Iranian and Libyan relations with European countries, following developments within the two countries targeted by the United States. Iraq has less to show; but still one cannot say that US–Iranian hostility, or US–Libyan hostility, is permanent, or that Iraq is a permanent pariah. In 2000 North Korea appeared to be beginning a process of normalization following the meeting of the two Korean presidents. Sanctions are often targeted against companies and do not cost the US taxpayer much. For the United States, the sanctions do not materially reduce the supply of energy, or increase its price in the domestic market; the immediate costs amount to little more than a few lost investment opportunities for US-based oil and service companies, and some acrimonious diplomacy with countries when sanctions are applied to non-US companies with operations in the US.

Long-term costs and effects of sanctions

It is likely that the United States will continue to use such sanctions – unilaterally if necessary, multilaterally if possible – to further its foreign policy objectives. The result is to put the security question elsewhere: how likely is an oil-exporting country to find itself in a position where its foreign or domestic policies provoke the United States to promote sanctions against it? This judgement is complicated because, under the US system of government, sanctions can be initiated by the Congress for domestic political reasons, and do not necessarily represent a careful calculation of national interest by the US administration.[19]

In the case of Libya at least, the sanctions may have contributed to changes in Libyan behaviour outside its borders. In the case of Iran, the question might be: how much better terms does Iran have to offer non-US companies to invest in Iran? The terms evident in the oil sector so far do not appear generous to foreign investors – indeed, it could be argued that they are actually more severe than those offered in the recent past by Venezuela or those available in a number of developing countries. This suggests that Iran, at least, does not need to incur major costs to avoid the effect of UN sanctions on its energy and energy-related trade.

Managing energy security risks

If the chances of success of a long-term oil cartel are discounted, concerns about energy security for economic, defence and foreign policy reasons are short- to medium-term questions. Subsidizing investment in extra permanent domestic energy supplies for normal use does not provide spare capacity when there are interruptions of the remaining trade. Interruptions can be taken care of by a combination of strategic stocks, collective measures such as the IEA emergency sharing procedures, and, of course, general foreign policy measures to pre-empt sanctions.

Strategic stocks are not negligible: stocks which cover 90 days of total oil consumption would cover 900 days of a 10% shortfall such as might occur if supplies from one country were disrupted (even Saudi Arabia supplied only 12% of world oil consumption in 1999). In the open and global market for oil and coal and the international markets for gas, it is almost always going to be cheaper to store fuel than to build spare capacity which is not used under normal conditions (the so-called spare oilfield idea).

[19] See Thomas Waelde, 'Legal Boundaries for Extraterritorial Sanctions', in Mitchell, ed., *Companies in a World of Conflict*, ch. 5, esp. pp. 189–92 (London: Royal Institute of International Affairs/Earthscan, 1997).

Economic security and self-sufficiency: conflicting objectives

Self-sufficiency in energy is a reasonable goal only if a country expects to be permanently at odds with the world (as South Africa did in the apartheid era) and expects to face widely applied sanctions as a result. This situation could arise for an energy-exporting country (such as Burma) as well as for an importing country. Elimination of energy trade is for most countries a prohibitively expensive option.

The question of economic security is a relative one: is a country more exposed than its main economic competitors and political rivals to the economic risks of energy trade? The answer is reassuring for Europe but not for Asia. In 1998 Europe and the United States both produced just over 40% of their oil consumption; Japan and South Korea produced none. Oil imports as a fraction of total energy requirements in 1998 were 23% for the United States and 24% for Europe; 51% and 56% for Japan and Korea respectively. Chinese oil imports are set to rise over the next two decades. In general, Europe is less exposed to the risks of energy imports than its principal economic competitors, and would damage its competitive position if it incurred high costs to reduce those risks.

The evidence of the last 25 years is that the growth of competitive markets, and the removal of price and investment controls, has led to the increased diversification of oil supplies and the development of substitute fuels. Countries that considered 'energy independence' options in the 1970s and 1980s have abandoned many initiatives because they were either too expensive or impractical, or both. The costs of subsidizing investment in domestic energy suppliers, or investing in subsidizing energy-saving, are such that few countries will choose to forgo the benefits of energy trade completely in order to avoid its economic risks. Capital-intensive investments in domestic energy – such as nuclear energy – would carry the same kind of commodity price risks as those in the competitive global markets. Even the United States, the world's richest and most powerful nation, abandoned the 'Project Independence' set up by President Nixon. For certain countries, such as Japan, independence is simply not feasible, though some otheres, such as France, have effectively secured their electricity system against the costs and benefits of trade by massive investments in nuclear power through a state-owned monopoly. The economics of nuclear power, quite apart from the environmental problems, do not seem to support such investments today. Moreover, nuclear energy and nuclear waste storage and treatment carry environmental and health as well as security risks.

Energy trade dependence and political security

The equation of 'security' with 'reducing import dependence' is dangerous for three reasons:

- In general, policies to reduce import dependence have to work within limits of affordability and acceptability which mean that they do not eliminate or even seriously reduce the risks attached to imports. Do changes of 5% or 10% alter a country's exposure to diplomatic and political pressures if the starting level is 50% or more?
- A similar argument applies to the question of how far oil exporters should incur high costs to diversify their dependence on oil exports. To deny the benefits of energy trade is generally expensive, and the more the economy is burdened with such costs the less competitive it will be. Weakening an economy for the sake of reducing energy trade also means weakening its capacity to invest in and sustain military defensive expenditure and support a proactive foreign policy.
- Finally, focusing on the level of imports in individual countries distracts attention from the global policies which can enhance energy security and national security for all countries by cross-border investment to increase global energy supplies and markets where it is cheapest to do so.

Energy strategies for security: a new prospectus

The object of these strategies is to promote economic and political security through the long-term economic development and diversification of energy trade globally, the efficient operation of international and national markets for fuels and fuel-related investments, and cooperation to mitigate the effect of short-term disruptions. A more specific agenda might look like the following:

- Promoting competition within domestic energy markets as well as internationally, to make substitution and inter-fuel competition stronger; this may involve reforming consumption taxes and subsidies which distort competition.
- Promoting free trade and investment in energy-related industries (and the transfer of energy-efficient technology) internationally. This includes the development of the Energy Charter Treaty to protect energy transit and cross-border energy investments.[20] It may also include a proactive role by international financial agencies in promoting the development of new cross-border pipeline infrastructures. What was done by the World Bank for Bolivia and Brazil, or the European Investment Bank for the line from Algeria through Morocco to Spain, may also be done for some of the Caspian and Central Asian export routes, or for lines for exporting east Siberian gas to East Asia.
- Addressing certain specific international political issues where there are implications for the expansion of international energy trade and investment. Examples are the status of the Caspian Sea and the South China Sea, where there are territorial disputes.
- Working with international companies and NGOs to develop acceptable codes of conduct regarding environmental protection and human rights in countries where energy developments involve problems of this kind.

What is different: international rather than national remedies

The policies described above, with their global focus, would have seemed unrealistic in the conditions of the early 1970s described earlier in this chapter. Now, they run with the trend of domestic and

[20] OECD/IEA, *World Energy Outlook 1999: Insights* (Paris, 1999): 'energy resources are significantly underpriced in eight of the largest countries outside the OECD, which represent collectively around one-quarter of world energy use' (p. 3).

international policies towards trade and security. They are cooperative in character, unlike the nationally centred policies which some of them contradict. There is a difference between designing policies which reduce the risks of international energy trade by making the flows of trade more flexible and more diverse, and policies which reduce a country's international energy trade absolutely. Promoting trade and investment, removing subsidies and rationalizing consumer taxes may go against national policies to subsidize and protect national energy industries.[21] Indicators such as 'import dependence' reflect only the old approach and do not reflect its costs under today's conditions, either directly to consumers or taxpayers or in terms of reducing the flexibility of international energy supplies and their ability to absorb shocks.

What continues: preparing for crisis management

Security policies are still needed to enable energy-importing and exporting countries to mitigate the effects on their defence capability, foreign policy and short-term economic activity of sudden but short-lived threats to supplies or markets.

For importers, the recipe is quite familiar:

- maintaining strategic stocks of fuels corresponding to essential demands;
- participating in collective schemes such as the IEA emergency sharing system in order to 'pool the risks'; there is clearly a case for trying to increase the participation of Asian countries in the IEA emergency response arrangements;
- being prepared, in extreme cases, to support UN actions to protect energy-supplying countries against aggression.

For energy exporters suddenly faced with sanctions which deny them markets, the available options are much less clear. It is another sign of

[21] For example, in the UK, by taxing energy conservation material and equipment for households more heavily than domestic fuels.

the general difficulties of achieving common cause among energy exporters that no collective response has been made to the sanctions applied to Libya or Iran. There is an imbalance of power against the exporters in this regard. But exporting governments need not be passive victims. The 'conventional wisdom' requires the continued expansion of oil production for export, especially from the Middle East. The timing of this expansion depends on the interests of the governments controlling the resources. It is also for them to determine the terms, if any, on which foreign investors will share in that development.

Summary

In contrast to 1973, we have today a world in which there is a 25-year record of expanding and diversifying oil and gas supplies in open and competitive markets; there is no Cold War and balance of superpowers; and governments almost everywhere are promoting competitive markets and fiscal discipline, rather than economic planning, as the keys to growth in a world of freer trade and cross-border investment.

In this world the economic risk of an OPEC or Middle Eastern cartel of oil exporters permanently distorting international oil prices in their favour is very small. The leading oil exporters are political rivals as well as economic competitors. Europe is less vulnerable to such a risk than its principal competitors; it has less exposure in terms of trade, and its foreign policies are not only less coherent but also less intrusive than those of the United States.

The economic element of energy trade risk can be further reduced by policies that promote the free movement of energy and energy-related goods and services (for example, by institutions such as the Energy Charter Treaty). Multilateral interest in the economic and political conditions for new cross-border infrastructure investment in the Caspian region and within Asia would also help to diversify energy supplies for Asia.

Policies to reduce the economic risk of energy trade by subsidizing domestic investments would limit the security and flexibility which the global system provides. Such subsidies and market distortions impose costs on the country which adopts them. Fluctuations in international oil, gas and coal prices are inevitable because of unexpected changes in demand, leads and lags, changes in investment conditions, and other factors which normally affect commodity prices.

The risk remains of temporary disruptions in supply or markets for political reasons. These may be accidental, caused by local instability or conflict, or they may be deliberate, as in the case of sanctions and embargoes. Since 1973 (and except for the case of South Africa), sanctions have generally been used by importers or the UN to deny markets to exporters, rather than to deny supplies to importers. Sanctions on energy imports or exports may threaten national independence, support foreign interference in domestic affairs, or restrain a country's foreign policy. There is therefore a case for governments to provide protection against such short-term shocks or threats. For most oil-importing countries, long-term policies to reduce oil imports or exports by subsidizing alternatives do not of themselves provide the flexibility to mitigate short-term shocks, because it is too expensive to give up the benefits of oil trade altogether, and because shocks and sanctions will be disruptive at any significant level of trade. Oil-exporting countries face parallel problems with regard to diversification from dependence on oil export markets. Their governments will look to take what advantage they can, however limited by competition, from their growing share of international trade in oil.

Expensive moves to reduce dependence on energy trade to avoid these political sanctions make sense only for a country which expects to be permanently in serious conflict with the world's major powers and with the UN. For other countries, the policies that best address these short-term risks are the same as those that were available in the 1970s: strategic stocks of oil and oil products, and emergency sharing schemes such as that of the IEA. The extension of such policies to more Asian countries is desirable because of their growing importance in the world economy and oil markets. Political and diplomatic attention to the causes of instability and sanctions would also help.

Chapter 8

Acceptability: stretching the limits

This chapter analyses the growing forces that challenge the acceptability of oil production and its use in transport, processing and heat and power, and speculates about their future trends.

The first section shows how challenges to acceptability have progressed. Challenges requiring local action led to national policies. Cross-boundary problems – marine pollution and the transport of hazardous waste – led to intergovernmental actions through classic international mechanisms including multilateral organizations with powers ranging from review and recommendation to dispute settlement.

The second section shows how global threats – holes in the ozone layer and climate change – have led to UN conventions which establish binding commitments and some international organizations which support them. The UN Framework Convention on Climate Change (UNFCCC), signed at Rio de Janeiro in 1992, and the Kyoto Protocol of 1997 face immense difficulties. They are intended to control emissions of greenhouse gases by billions of users; the allowances are linked to a 1990 baseline which is out of step with current levels of OECD economic activity. There are no quantified commitments for developing countries, which will soon account for more than one-half of global GHG emissions. The treaty envisaged that commitments by the developed countries should precede negotiation of any more specific commitments by developing countries. Effective climate change mitigation policy will need more and different policies. Their impact on oil will depend on the balance between targeting carbon and spreading the cost of change among users of different fuels.

The third section puts the 'acceptability' of oil in the context of sustainable development and the emergence of 'private international relations' facilitated by the communications revolution and the expanding effectiveness of transnational non-governmental organizations (NGOs) in promoting ethical values. Investors in private sector companies are concerned about reactions by 'civil society' to companies' activities. The issues are not limited to environmental matters, or even to human sustainable development objectives, but reach through human rights to conflicting political values. The gap between countries with democratic legitimacy and honest, competent governments and those less favoured, is likely to become better defined and better targeted by those seeking worldwide commitment not only to sustainability and development but also to democracy and social justice.

The implications for the oil sector are strong and are not limited to the possible effects on the world demand for oil. Intergovernmental action and the pressures of opinion through 'private international relations' will apply new definitions of social acceptability for international access to many actual and potential oil-producing countries for exploration, production and transport infrastructure. These challenges may extend to the structure of governments and the processes by which they claim legitimacy. The promotion of 'good government' abroad as well as at home may become a stronger imperative among those committed to international markets for goods, services and investment, and the free flow of ideas about how to achieve a better life.

Acceptability begins at home

The classic problems of environment, health and safety (EHS)

Concerns about health and safety have prompted a great part of the environmental responsibility which companies accept or have imposed upon them. In many companies 'environment, health and safety' (EHS) has been organized as a discrete department, with designated executive responsibilities, and headlined as a single policy. EHS issues are typically local, not necessarily national (except when regulation or litigation makes them so) and rarely international in effect. Such issues may be informed by the international spread of knowledge, for example, about the causes of accidents or the effects of toxic substances. Mitigating the risks may be made easier by the international spread of technology and management practices for care and caution, such as the International Organization for Standardisation (ISO) codes. Multinational companies can spread the practice of their 'best' operations to places where local conditions and regulations might allow something less. Nevertheless, the international dimension is not normally a key part of the problem – though it may be a part of solutions. The problem is to avoid damage or reduce risks in a particular place. The *result* is generally local, even if the knowledge and discipline have been drawn from a global pool. Such issues, important though they are for the business concerned and its customers, contribute to international relations only by exception.

Exceptions might occur when, for example, a developing country fails to protect intellectual property rights and deters multinationals from transferring technology which would improve the environment. EHS legislation, such as the US Clean Air Act of 1980, may be involved in a way that constitutes a restraint of trade or trade discrimination under GATT rules, as in the 1995 complaint by Venezuela against US stipulations for imports of gasoline.[1] Such cases have an international dimension, but the international factors are dealt with through existing international procedures established within a much wider framework of international relations.

Local results may cross frontiers. Prevailing winds may blow clouds containing sulphate aerosols across borders (from the UK to Norway, or from north-east China to Japan), for example, to fall as acid rain on valuable forests. Typically, such problems have entered international relations through classic channels of diplomacy and international negotiation between the countries concerned.[2] Such negotiations have not always been successful, since the interests of the polluting and the polluted parties are opposed on some points. However, neighbours are neighbours and have continuing common interests in avoiding common or reciprocal environmental damage. There has been a steady

[1] The case arose because the United States applied different rules on the chemical characteristics of imported gasoline from those for domestically refined gasoline. The WTO disputes panel ruled in 1996 that the discrimination conflicted with WTO obligations to apply the same restrictions to domestic products as to imports: the US was obliged to end the discrimination (but not the restriction on gasoline components). See outline in <www.wto.org/disputes/>.

[2] The 1979 Stockholm Convention on Long-range Transboundary Air Pollution is the principal example.

growth of regional agreements aimed at solving, for example, the problems of pollution in areas such as the North Sea and the Mediterranean.[3] Countries not affected could neither contribute to nor impede the regional solution except by way of sharing science and technology through commercial and academic channels. Countries outside the region may play a supporting role where the regional problem is connected with the long-distance transit of oil or gas, as in the Black Sea for Caspian exports, or the Straits of Molucca for Middle East–East Asian trade.

Pollution at sea

The case of oil spilt at sea is different. International waters, variously defined, have long been outside national jurisdiction: ships are under the jurisdiction of the country in which they are registered. Spills, or deliberate dumping of waste (such as oil residues) may drift into national waters and coasts as well as damaging resources in international waters, such as fish stocks, in which countries have economic interests. In this context the problem was deemed to arise not so much from the trade in the goods involved, but rather from their careless or unsafe handling while in transit. The carriers, rather than the goods, were the object of sanction. Once landed, goods were treated the same regardless of how they were carried: trading rights under GATT and the WTO were not modified. No one country could effectively protect its own interests by unilateral action. Careful behaviour by its ship-owners could be useless if ship-owners from other countries were careless or wilful in their treatment of the marine environment.

Global measures to reduce the risk of marine accidents, and pollution, were developed through painfully negotiated international conventions. The convention of 1948 establishing the International Maritime Organization (IMO), which eventually came into operation in 1959, built on previous conventions concerned mainly with safety of crews and passengers. New international conventions (binding on states) and codes

[3] See Tony Brenton, *The Greening of Machiavelli* (London: Royal Institute of International Affairs, 1994), pp. 95–100.

(with the status of recommendations) have been developed to cover a wide variety of topics affecting the transport of goods and passengers by sea. The 1954 International Convention for the Prevention of Pollution of the Sea by Oil (OILPOL) became the basis of a series of conventions and codes of practice.[4] These conventions, supported by IMO codes of conduct, were responsible for a number of profound changes in the operations of seaborne transportation of oil. The dumping of oil at sea was outlawed. Vessels are required to retain and segregate tank washings (mixtures of oil and water) for delivery to separation facilities at the port of destination. The use of inert gas blankets, segregation of ballast tanks, and double hulls has been mandated.

As well as prevention, international agreements offer cures and sanctions. Through the International Convention on Oil Pollution Preparedness, Response and Cooperation (OPRC),[5] the IMO provided for a global system for coordinated response to oil spills. A 1971 convention established an International Fund for Compensation for Oil Pollution Damage.[6] This fund, financed by oil importers, provided a back-up to the funds available from ship-owners to meet civil liabilities for pollution damage. As well as setting standards for ship design and operations, the IMO is also the foundation of a code on Standards of Training Certification.[7] In relation to these standards – critical to the safe operation of vessels – the IMO has review authority over member states' compliance; IMO member states have the right to inspect ships entering their ports and to deny access to substandard ships.[8] International action under these conventions has significantly altered the legal framework for the transportation of oil by sea and the operational practice of the shipping and oil industries.

---

[4] OILPOL was amended in 1969 following the *Torrey Canyon* disaster; a further Convention on the Prevention of Pollution from Ships was adopted in 1973: its reach was greatly expanded by the 1978 Marine Pollution (MARPOL) Protocol (effective from 1983) and subsequent protocols.

[5] Adopted 1990, in force 1994.

[6] Adopted 1971, in force 1978.

[7] Adopted 1978, in force 1984.

[8] An outline of the development of the IMO and related conventions is available on the Internet at <http//www.imo.org/imo/50ann/history>.

Three characteristics of the development of marine anti-pollution regimes are of possible interest to those concerned with developing international regimes for climate change policies. The current marine conventions took decades of negotiation. The process has been speeded up since 1974 by the adoption of a 'tacit consent' procedure by which new protocols take effect after a set date unless rejected by one-third of the contracting parties representing 50% of the combined tonnage covered by the parties to the agreement. The IMO agreements, protocols and codes integrate anti-pollution provisions with measures dealing with safety, standards, operations, liabilities, and the development of regulatory capacity in developing countries. Enforcement ultimately depends on a national government or government agency taking action against the ships registered under its flag or calling at its ports.

The UN Convention on the Law of the Sea (UNCLOS) also dealt with protection of the marine environment within the framework of a much wider convention. UNCLOS covers the management of fish stocks, navigation rights, the definition of territorial waters and rights over the continental shelf and coastal waters, settlement of disputes (including a new international tribunal), and an International Seabed Authority to regulate mining of the deep seabed. UNCLOS reinforces the IMO-related apparatus, adds to obligations to prevent pollution of the sea from land-based sources (not covered by the IMO protocols), and clarifies rights of coastal states to apply environmental protection in their territorial waters and 'exclusive economic zones' (EEZs: generally, areas up to 200 nautical miles offshore). It also clarifies the rights of receiving (port) states to enforce internationally agreed rules on vessels voluntarily entering their ports (except with regard to violations in the territorial waters of the state where the ship is registered). Together with the inspection procedures and standards provided under IMO, this provides a powerful tool for applying standards to ships of flag states that are not parties to IMO or that apply IMO procedures in a lax or ineffective manner.

UNCLOS contains a notorious example of a false start in international treaty-making. This may be a warning to climate change treaty-makers, although it has no direct relevance to energy and climate issues. The provisions for international control of deep-sea mining were, for decades, contentious between groups of industrial and developing countries. Eventually, they were amended by agreement in 1994 to exclude earlier provisions for the establishment of an 'international enterprise' to carry out deep-sea mining and control production levels, mandatory acquisition of technology and appropriation of profits for international distribution. The 'international community' was unwilling to use the treaty to redistribute wealth. The 1994 amendments also incorporated 'GATT language' to ensure that seabed mining is conducted within the rules of GATT.

Dangerous goods

The IMO adopted in 1965 a Code on the Maritime Transport of Dangerous Goods (IMDG). This did not have the force of a convention and applied only to movements by sea. The UN Convention on the Movement of Hazardous Waste – the 'Basel Convention' of 1989[9] – was designed to halt or at least minimize the movement by air, land or sea of hazardous waste from countries where its disposal was strictly controlled and costly to countries where it would be disposed of in a manner threatening to human health and the environment. The convention places trade in hazardous waste under strict controls in the exporting (mainly OECD) countries: the principle, broadly, is that exports of hazardous waste must be notified to competent authorities in the importing countries and is illegal unless the latter consent.[10] There is agreement that such exports should be banned to countries (such as developing countries) lacking the technical or administrative capacity to manage such a system. The question of trade in goods for recycling is still under debate.

The Basel Convention has features which may be relevant to the emerging climate change mitigation regime: a global solution is

---

[9] In effect, 1992. For an outline, see Jonathan Kreuger, *International Trade and the Basel Convention* (London: Royal Institute of International Affairs, 1999).

[10] This may be regarded as a weak gesture towards Principle 21 of the Stockholm Convention which requires states to prevent, if they have the necessary control, activities that damage the environment of other states. Principle 21 couples this with a restatement of states' sovereign rights to determine their own environmental laws.

necessary but the obligations undertaken by developed countries are different from (and more severe than) those of developing countries. The operational provisions apply to transboundary movements. The convention looks in theory at the whole life-cycle of the product and seeks to minimize the generation of hazardous waste. Compliance (or violation) occurs when boundaries are crossed. There is believed to be a large illegal trade in hazardous waste, despite the provisions of the convention. There is potential for conflict between enforcement of the convention and respecting trading rights protected by GATT. In this last aspect the Basel Convention is representative of a class of multilateral environmental agreements (MEAs).[11] There are now over 200 MEAs.[12]

The world pushing at the door

During the 1970s a new class of problem climbed the agendas of governments and businesses, prompted both by better scientific understanding and by pressure from organizations of people concerned about wider environmental issues. These were problems of environmental damage which had global results. The appearance of growing 'holes' in the ozone layer, the loss of entire species from the planet, and the phenomenon of climate change cannot be avoided by action in one country. It is the global result that matters. Actions in individual countries may make important differences, but wide international cooperation is the only way to achieve the global result. Key examples of such cooperation are the Montreal Protocol on Substances that Deplete the Ozone Layer (1987) and the Convention on International Trade in Endangered Species of Wild Flora and Fauna (CITES).

---

[11] For a discussion see Duncan Brack, 'Multilateral Environmental Agreements: An Overview', in Halina Ward and Duncan Brack, eds, *Trade, Investment and the Environment* (London: Royal Institute of International Affairs/Earthscan, 2000).

[12] The ten major environmental conventions are summarized in UNEP, *Global Environmental Outlook, 2000* (London: UNEP/Earthscsan, 2000), pp. 199–206.

The ozone layer

The Montreal Protocol has no direct application to the oil industry, but is important as a precursor to the UNFCCC.[13] Concern about the ozone layer, like concern about climate change, originated in scientific observations and prediction rather than events immediately damaging to health or the visible environment.[14] Public perception of the threat posed by the 'hole in the ozone layer' reinforced official concern. The 'precautionary principle' was invoked to justify action even though the science involved was not fully understood at the time and it was not clear how easy or cheap it would be to replace chlorofluorocarbons (CFCs). International scientific studies, and competitive ideas, eventually convinced manufacturers of ozone-depleting substances (ODS) such as CFCs to work with governments in negotiating feasible phase-outs, the supply of alternatives, and acceptable ways of meeting the needs of developing countries that import these substances. To provide incentives for countries to sign up to the protocol, and to prevent 'leakage' of ODS business to non-participant countries, the Montreal Protocol provided for member countries to ban imports not only of ODS but also of products containing ODS. These sanctions, rather than the small funds available for developing alternatives to ODS, seem to have been the main factor in securing wide participation by developing countries in the protocol,[15] and therefore its apparent success. The problem of illegal trade nevertheless exists.

As with many MEAs, there was potential for conflict with GATT provisions, especially during the period when many developing countries, members of GATT, were not parties to the protocol and might therefore have been subjected to trade sanctions. GATT Article XX (providing for general exceptions) in para (j) allows discrimination by members of commodity agreements against non-members in order to fulfil the

---

[13] The term 'Montreal Protocol' is used here to refer to the Vienna Convention for the Protection of the Ozone Layer, 1985 and the Montreal Protocol on Substances that Deplete the Ozone Layer, 1987.

[14] For an account of the origins of the Montreal Protocol see Duncan Brack, *International Trade and the Montreal Protocol* (London: Royal Institute of International Affairs/ Earthscan, 1996), ch. 2, pp. 10–26.

[15] See ibid., pp. 51–9.

objectives of the agreements, provided always that the commodity agreements meet certain pre-specified criteria. There is no similar exception for action against non-members of an MEA, though there is a general exception for action to protect 'human or animal or plant life and health' (para. (b)) and for the conservation of exhaustible resources (para.(g), later extended to include living resources such as fish). There is deep controversy over the question of whether GATT should be amended to allow MEAs to use sanctions against non-MEA members in order to fulfil MEA objectives. The failure of the Seattle WTO meeting to initiate a new round of WTO negotiations has left this issue on the table.

Climate change

There are no feasible ways of preventing climate change in one country. Global warming is truly global. To the extent that it can be mitigated by human actions, those actions need to be taken by a high proportion of those whose actions release GHGs and may therefore threaten the global climate. In the similar case of ODS, the Montreal Protocol aimed to phase out the production of ODS as a way of eliminating their use. Actions by individual countries were reinforced by trade sanctions and assistance to developing countries needing alternatives. This 'supply-side' global approach has not been followed in the UNFCCC and the Kyoto Protocol.[16] Oil currently accounts for over 40% of global carbon emissions, which, in turn, form around 80% of global GHG emissions.[17] However, the countries committed to reducing CO_2 emissions (Annex I of UNFCCC[18]) account for only

[16] For an analysis of the evolution of the Earth Summit agreements in terms of the substance of the issues and the international relations processes involved, see Brenton, *The Greening of Machiavelli*; for the Kyoto Protocol, see Michael Grubb, *The Kyoto Protocol: A Guide and Assessment* (London: Royal Institute of International Affairs/Earthscan, 1999).
[17] Aggregation of the global warming potential of different GHGs is complex.
[18] Essentially, 'Annex I' means the OECD, except Mexico and Korea, and Russian and related economies in transition. There are small differences between the 'Annex I' list of UNFCCC and the 'Annex B' list of the Kyoto Protocol, which excludes Turkey and Belarus and includes some other small east/central European countries.

40% of oil production.[19] This is different from the situation in respect of substances affected by the Montreal Protocol, where 90% of the production in the base year took place in countries committed to phase out ODS (the non-Article 5 countries). An 'upstream' policy of reducing CO_2 emissions from Annex I countries by reducing the supplies of fossil fuels has never been put on the intergovernmental agenda.[20] Any supply constraint would have to apply to imports as well as domestic production. Annex I countries consume about 65% of the world's fossil fuels and 68% of the world's oil.

It is arguable that policies to reduce supplies of oil and gas would be relatively easy to implement within the OECD: licensing of new developments, production quotas, and corresponding restrictions on imports – as allowed by GATT – have been carried out before on oil in the United States (for other reasons). The scope for expanding oil and gas production in the OECD is in any case limited. In the case of Russia and the other economies in transition, the production of all fossil fuels is well below 1990 levels, though it could grow (the same anomaly as exists for fossil fuel consumption). The problem would be to restrict the scope for expanding the production of coal in countries like the United States, Canada and Australia. This would have to be done either by means of innovative production quotas, or by direct restrictions on the use of coal in its principal market – the power sector – as is currently implied in the Kyoto mechanisms.

If supplies of fossil fuels (domestic production and imports into the group as a whole) were restricted in the Annex I countries as a group, fossil fuel prices within the group would rise relative to other fuels and to the cost of fuel-avoiding technology.[21] 'Windfall' profits for domestic

[19] The Annex I countries' share of production of other fossil fuels is higher than for oil: just over 70% for natural gas and about 50% for coal. Annex I countries at present account for nearly 70% of world fossil fuel consumption and nearly the same proportion of world fossil fuel imports.

[20] The Greenpeace 'carbon logic' campaign to obstruct new oil developments in North America and the North Atlantic implicitly leans in this direction. See Bill Hare, *Fossil Fuels and Climate Protection: The Carbon Logic* (Amsterdam: Greenpeace International, 1997).

[21] The final effects would be complicated, depending on the severity of the controls. If the fall in demand led to lower fossil fuel supply prices, the price effect might be offset, but the physical control would be binding.

fossil fuel producers could be captured (as they were following earlier oil price shocks to importers) by taxing governments or by the sanctioning of supply rights. These new government revenues could be recycled (as is suggested for carbon taxes) to reduce other, inefficient taxes – such as payroll taxes – and offset the economic damage which the rising price of scarce fossil fuels would cause. The prospect of high prices within the 'supply reduction' countries could attract fossil fuel exporters to join: the higher prices would compensate them, at least in part, for revenue lost by the lower export volumes.

Supply restrictions which raised fossil fuel prices would provide general signals to encourage economy in the use of fossil fuels similar to those sent by an 'upstream' carbon tax. Allocating and adjusting quotas for suppliers would be a political process; but it would be a simpler process, with fewer targets (and corporate rather than consumer targets) than may be required in the systems for emissions allowances which will be necessary to make the Kyoto Protocol work. Compliance with production and import allowances would be easy to measure, though an obsolete baseline for supply restraint, like a demand restraint, would create 'winners and losers'.

Why was a supply-side approach not considered? The obvious reason is that it is emissions from fossil fuel use that are the problem; expanding production together with the use of sequestration or 'carbon sinks' might be part of the solution. Another reason might be the hostility shown by domestic fossil fuel producing interests and oil and coal exporters to the whole idea of climate change policy. Their opposition applies to demand-side measures also. They may have missed a chance of getting a supply-side policy they could understand, influence and work with.

A further, broader, political reason might be that the momentum for climate change policy drew on growing public concerns about visible pollution. There was a perception of the environment as an exhaustible resource. The economics of the problem of 'the commons' suggested that state action was required to correct the operation of the market. Intervention on the demand side could reopen opportunities for social engineering and reform (or political patronage) which depend on active intervention by the state in everyday life. The argument between 'command and control' regulations and 'economic instruments' like general, simple carbon taxes, may be more than technical: it can draw on general prejudices about the possible and desirable role of the state in organizing the detailed life of society.

A more direct explanation could be that policies of reducing supplies of oil (and thereby increasing its price) would be inexplicably contrary to the traditional policy of 'securing' oil supplies at 'reasonable' prices. The reasons why supply reduction seems a perverse way of reducing CO_2 emissions may be a clue to why the policies of demand reduction to which Annex I countries are committed are proving so difficult to realize.

Kyoto: climbing the foothills

Since the UNFCCC Convention was signed and came into force there has been a steady progression of conferences of the parties to that convention seeking to clarify, strengthen and extend its objectives.[22] The UNFCCC itself was a step on a path whose earlier milestones were the Stockholm Conference on the Environment in 1972 and the Brundtland Report, *Our Common Future*, in 1987. Of these, the Kyoto Protocol to the UNFCCC is currently the most important.[23] If ratified, it will bind the Annex I parties to reduce or contain emissions of GHGs to amounts assigned in relation to a baseline of 1990. The amounts become binding limits during the 'commitment period' of 2008–12. Non-Annex I parties – essentially, the developing world – have no such commitments under Kyoto, though they have generalized and vague obligations under both the UNFCCC and the Kyoto Protocol.

[22] See Brenton, *The Greening of Machiavelli*, for a description of the process and an analysis of the relations between the evolving 'Rio' process: and other environmental agreements and developments in international relations (such as North–South dialogue) during this period.

[23] For a short guide, see Grubb, *The Kyoto Protocol* (London: Royal Institute of International Affairs/Earthscan, 1999). There are also short summaries in John Weyant and Jennifer Hill, 'Introduction and Overview', in *The Cost of the Kyoto Protocol: A Multi-Model Evaluation*, special issue of *Energy Journal*, ed. John P. Weyant, and World Energy Council, *The International Environmental Agenda* (London: Energy for Tomorrow's World, 2000), pp. 37–43.

The UNFCCC, the Kyoto Protocol, and the remaining climate change policy agenda have been extensively analysed. The International Panel on Climate Change (IPCC) continues to produce a series of specialist reports analysing scientific and economic issues concerning climate change and climate change mitigation policies.

Box 8.1: Intergovernmental Panel on Climate Change

The IPCC was established in 1988 by the World Meteorological Organization and the United Nations Environment Programme:

- to assess available information on the science, the impacts, the economics of, and the options for mitigating and/or adapting to climate change; and
- to provide, on request, scientific/technical/socio-economic advice to the contracting parties to the UNFCCC.

The IPCC works through three working groups:

I deals with assessing scientific information on climate change;
II assesses the environmental and socio-economic impact of climate change; and
III formulates options for national and international responses.

The first 'scientific assessment' of the IPCC was published in 1990, and was followed by supplementary and working group reports which were a major input to the 1992 Earth Summit at Rio de Janeiro which led to the formation of the UNFCCC. The Second Assessment Report of the IPCC (1995) was a major input to the conference which drew up the Kyoto Protocol in 1997. A third report is being completed during 2000–1 in preparation for the 'Rio + 10' follow-up to the Earth Summit, planned for 2002.

The working groups are staffed by experts and draw on an immense worldwide network of academic and specialist experts. The final IPCC exports are the product of agreement by the government nominees.

Tying the ropes

Certain problems exist because of the way in which the convention and protocol were structured and the implicit political 'deals' that tie the countries' commitments together. The choice of demand-targeted rather than supply-targeted policies (discussed above) was built into the process before UNFCCC and was not in question. Once it was recognized that policies would target fossil fuel use and GHG emissions rather than supply, the question of allocating the right to use – or rather to emit – became critical. Any decision to base emission rights (the 'assigned amounts') on anything other than an historic baseline period would have opened the door to irreconcilable struggles for new principles of allocation which would redistribute economic rights among countries. In 1992, 1990 – the latest year for which consistent figures were available – did not seem an abnormal baseline: few could have foreseen the speed and depth of collapse of the formerly centrally planned economies of the FSU and eastern Europe.

The UNFCCC (Annex I) commitment to 'voluntary actions' to reduce emissions to 1990 levels looked equitable but was necessarily inefficient. Countries are differently able to change fossil fuel consumption, depending on their non-fossil resources, the cost of change, and the degree to which energy efficiency is already promoted by tax and regulatory policies.

Differences in existing energy taxes would have made even the idea of a new, universal (Annex I) carbon tax 'unfair'. For countries with low initial tax levels, adding a common carbon tax would have had greater economic impact (and greater impact on energy prices) than in countries where energy taxes ware already high, creating an 'unfair' allocation of the burden of change. This problem of sharing the burden in an acceptable way applied also to the academically attractive idea of reforming existing energy taxes to put them on a carbon basis.[24] Moreover, the economies, and therefore the energy demands, of even Annex I countries were growing at different speeds: commitment to return to 1990 levels creates fewer problems for slow growers than for fast growers. Such problems inevitably meant that some differentiation of commitments had to be negotiated; and the Kyoto commitments were the result of this acknowledgment. Yet even with the differentiated commitments established at Kyoto, countries (and people within countries) face different opportunities for adjustment. The various trading and flexibility regimes offer ways to find global least-cost solutions, but the allocation of the initial commitments nevertheless still creates winners and losers. The size of the gains from emissions trading are a measure of the size of the redistribution of adjustment costs.

[24] OECD Economics and Statistics Department, *Energy Prices, Taxes and CO_2*, Working Paper 106 (Paris: OECD, 1991).

What UNFCCC–Kyoto requires is a change in the use of fossil fuels similar to that which occurred in the 1980s following the oil price shocks of the 1970s. The 1992 UNFCCC commitment required voluntary policies from Annex I countries individually to stabilize CO_2 emissions in 2000 at their 1990 levels (a commitment which is not being met in most OECD countries). The Kyoto Protocol, if ratified, will require individually specified percentage reductions by Annex I countries (in all GHGs) to be achieved during the period 2008–12. Together, the Annex I reductions should bring emissions down to at least 5% below the 1990 level – or about 12% below the 'conventional vision' reference case for 2010.[25] Great care has been taken to avoid describing the quantities which result from applying the reductions to the 1990 baseline by terms like 'entitlements', 'allowables' or 'rations', though this is what they are (the term used is 'assigned amount'). A country may choose whether to translate its national 'assigned amounts' into individual allowances or to rely on policies which it believes will limit emissions to the national total, or some combination thereof. Countries may take different approaches to the energy-intensive industries which face competition in export markets. This opens the possible need for 'border adjustments', i.e. tariffs on imports – which may conflict with GATT/WTO obligations.

Clouds on the trail

The Kyoto Protocol therefore allows several different forms of flexibility to supplement domestic actions. The details of the flexibility mechanisms remain to be defined at the Hague Conference of the Parties in 2000. Broadly, these are as follows:

- *Trading 'assigned amounts'*. A country can 'buy' the part of the allowed emissions quota of another country. The US sulphur dioxide permit trading system set up in 1995 was a success in that trades occurred far below the prices expected before the scheme was set up.[26]

[25] See EIA, *International Energy Outlook 2000*, table 2, p. 13.

[26] Mainly for the unrelated reason that the deregulation of the railways led to a reduction in the cost of delivering low-sulphur coal from the western to central and eastern United

Whether or not cross-border trading systems are developed, there is nothing in the protocol to prevent a country allowing trading between corporations within its borders. Some corporations in the UK are experimenting with internal 'trading schemes' to reduce the cost of meeting self-imposed CO_2 emissions reduction targets. In 1997 it was clear that the formerly centrally planned economies were not soon going to recover and that their likely 'business as usual' GHG emissions would remain below 1990 levels for some decades. Involuntarily, a new resource had been created – 'hot air' – to add to any trading potential created by real changes in the use of fossil fuels in the Annex I countries which continue to grow their economies. By continuing to focus on an outcome for Annex I as a whole, other economies which had grown could 'buy' assigned amounts of emissions from countries which were unlikely to need them.

- '*Bubbles*'. The EU, on ratification of the protocol by its member states, achieves recognition of a 'bubble': that is, it can agree a joint emission reduction (to be declared on ratification), and so long as this is met some member countries may emit more than their individual emissions allowances.
- *Joint Implementation* (JI). This is essentially the bubble concept related to a project or programme, and the outcome is similar to trading: the country where a cheap emissions reduction project is feasible meets some of the reduction target of another country.
- *The Clean Development Mechanism (CDM)*. This essentially enables an Annex I country to meet its emission reductions target through funding and transferring technology to a project in a non-Annex I country where a verifiable reduction in emissions from 'business as usual' is the result.
- *The GHG basket*. The reduction targets are for the basket GHGs as a whole: allocation among the various GHGs is at the discretion of the country concerned.

---

[26] (cont)
… States. See Denny A. Ellerman et al., *Emissions Trading Under the US Acid Rain Program: Evaluation of Compliance Costs and Allowance Market Performance* (Cambridge, Mass.: MIT Center for Energy and Environmental Policy, 1997).

- *Sinks and sequestration.* Under conditions not yet well defined, changes in agricultural or land use practice which would not otherwise have occurred, and which lead to an increase in CO_2 absorption, may lead to changes in a country's assigned amount. It is possible that cheap technology for injecting CO_2 into depleted oil and gas reservoirs and coal mines could provide a major 'sink' disposing of CO_2 from large chemical/power/refinery plants in the future.
- *Crediting and banking.* Under conditions not yet fully defined, JI and CDM projects that achieve sustainable emissions reductions before 2008 can credit those reductions to the 2008–12 period. The protocol also provides generally that a country whose emissions fall below their assigned amounts in the commitment period 2008–12 may carry the balance forward as a credit to its assigned amounts (yet to be negotiated) in subsequent commitment periods.

Much of the economic modelling work on the effects of meeting the Kyoto commitments has concentrated on demonstrating the economic benefits of emissions trading, on quantifying the long-run effects of meeting the Kyoto targets, or on extending them beyond the period 2008–12.[27]

The shifting landscape

At Kyoto in 1997 it was not known that the US economy would continue to outpace European economic growth, and Japanese economic recovery would remain delayed, with the effect that the 1990 baseline now appears a serious distortion (Table 8.1).

The combination of divergence in growth between the OECD and the former centrally planned economies, and divergent growth among OECD members, has led to acute differences in the degree of change which would be required for Annex I countries to meet their commitments not to exceed their allocated amounts. Table 8.2 shows the

[27] See Weyant, ed, *The Cost of the Kyoto Protocol*. Only 1 of the 13 modelling teams (Oxford Economic Forecasting Ltd) contributing to the Energy Modelling Forum Exercise reported here was essentially macroeconomic in scope.

Table 8.1: Increase in real GDP since 1990

Country	% increase
USA	32
EU	23
Japan	12

estimate provided by the US Department of Energy for CO_2 (the GHG 'basket' would allow some flexibility in these numbers). If the possibilities of 'trading' are excluded, the change from the reference case for the OECD countries would range from 30% for the US to 22% for Japan and 15% for western Europe. The US government insists that the reduction targets agreed at Kyoto took the possibility of trading into account, but the EU argues that such trading should be limited so that each country achieves 50% of its target reduction within its own borders – even if cheaper reduction can be achieved elsewhere through trading, JI or CDM projects.

Table 8.2: Kyoto CO_2 commitments versus conventional vision baseline

	Kyoto 'allocated amounts' 2008–12	
	% change from 1990	% change from 'conventional vision' projection
USA	−7	−30
Canada	−6	−30
Western Europe	−8	−15
Japan	−6	−22
Australasia	7	−23
FSU	0	44
Eastern Europe (Annex I)	7	23
Total Annex I	−4	−12

Source: EIA, *International Energy Outlook 2000*, table 2, p. 13.

The US end of the rope

These numbers clearly show the US interest in maximizing flexibility and avoiding any restrictions on the trading of assigned amounts. Buying 'allocated amounts' from Russia or elsewhere may be cheaper than adjusting US fossil fuel consumption at the margin,[28] but the adjustment required from the United States would still be proportionately far greater than that required from slower-growing Europe. To some extent, this is inevitable. The United States is the largest economy in the world, with a large share of world GDP and very high levels of GDP and emissions per capita. Unless there are special climatic factors in its favour, the US economy and the lifestyle of its citizens are at risk from the consequences of climate change. Most cost–benefit methods of valuing damage take account of the wealth and income of the person damaged: damage to an average US home will cost more to repair than damage to an average home in Bangladesh. The United States may therefore capture a significant share of the economic benefits of mitigating climate change and can reasonably be expected to bear a large portion of its costs. However, most studies suggest that the United States will probably suffer fewer damaging effects from climate change than some developing countries. Bangladeshi homes are more likely to be flooded than US homes, for example. This reinforces the idea that the US share of adjustment costs implicit in the Kyoto Protocol would be disproportionately heavy.[29]

This simple view of public and private interest is a strong reason why the United States may not ratify the Kyoto Protocol (whatever the state of US public sentiment about climate change) or, if it does, may look for the maximum use of the flexibility provisions. The United States may be expected to resist any attempt to limit its purchase of allocated amounts from Russia and similarly placed economies in transition, and to seek to introduce some mechanisms for capping the price of trades in allocated amounts of GHGs.

---

[28] As Russia will dominate the supply of excess 'assigned amounts' it might have the opportunity to exercise market power over pricing these sales. It would be rational to price as close as possible to the costs avoided by buyers, which in the US case might be quite high. Most of the economic benefits of trading would then be captured by Russia, rather than by purchasers such as the United States. See Ulritsch Bartsch and Benito Müller with Asbjørn Aaheim, *Fossil Fuels in a Changing Climate* (Oxford: Oxford University Press/ Oxford Institute for Energy Studies, 2000), p. 182.

[29] See William Nordhaus and Joseph Boyer, *Requiem for Kyoto: An Economic Analysis*, special issue of *Energy Journal*, 1999.

Rocks above

The degree to which 'trading' or similar flexibility involves the Annex I economies in transition thus has an important effect on the degree of change required of the other Annex I countries. The commitment to an aggregate 5% reduction from 1990 levels implies an aggregate reduction of 12% below the conventional vision for 2010 only if all commitments are aggregated through the 'trading' of allowable emissions.

Even with flexibility, if the changes are introduced quickly (as the Kyoto targets for 2008–12 imply) the potential effects of a 12% reduction can be compared with those of the oil price shocks. The fall in OECD oil consumption from 1973 to 1983, following the two oil shocks, was 13%. Something has been learned about macroeconomic policy in the face of energy price shocks, and energy, labour and capital markets are in most industrial countries more flexible now than they were in the 1970s. Nevertheless, the list of problems would be the same. Consumers' fuel-using equipment – factories, vehicles, buildings – will be prematurely obsolete or suffer a loss of income-generating capacity. Large sums of money will be removed from consumers of fossil fuels and redistributed to those whom the political process in each country favours, creating uncertain winners and losers from revenue recycling. The inflationary effect of higher costs in relatively rigid economies will lead either to actual inflation or to tight money policies which will depress growth.[30]

In an equation where full flexibility means managing the equivalent of the 1970s oil shocks, it is unlikely that lesser flexibility – and greater shocks in individual countries – will be preferred by most countries except as part of the politics of gesture. Loss of flexibility would hurt the United States most (for reasons given below). The chance of the

---

[30] See a discussion of macroeconomic effects in Adrian Cooper, Scott Livermore, Vanessa Rossi, Alan Wilson and John Walker, 'The Economic Implications of Reducing Carbon Emissions', in Weyant, ed., *The Cost of the Kyoto Protocol*. For a carefully modelled example for the United States, see Charles Rivers Associates and DRI/McGraw Hill, *Economic Impacts of Carbon Taxes: Overview* (Palo Alto, Ca., Electric Power Research Institute, 1994).

United States consenting to serious restrictions on the flexibility mechanisms agreed at Kyoto is therefore low.

The disappearing horizon

Annex I countries were responsible for two-thirds of the world's carbon emissions in 1990. Under the 'conventional vision' they will generate less than 50% of world emissions by 2020.[31] It would be several further decades (depending on economic growth rates) before the Annex I countries' share of the accumulated concentrations of GHGs in the atmosphere fell below 50%.

It is clear that the growth of GHG concentrations cannot be halted by any plausible reduction in Annex I emissions alone. Between now and 2020 Annex I countries would have to reduce emissions to match non-Annex I countries' increases, and as the Annex I share of GHG emissions fell, the proportionate reduction required of them would rise – to the point where all emissions came from non-Annex I countries. If the rise of GHG concentrations over the next half-century is a threat to the global climate, it cannot be seriously reduced without both Annex I and non-Annex I countries reducing their GHG emissions during this period.

Without non-Annex I participation, the Annex I commitments are at worst symbolic, at best experimental. Securing non-Annex I participation in the assignment of GHG allowable emissions requires some move away from the idea of a baseline to the allocation of future emissions rights (within a constraint of acceptable concentrations) according to principles which non-Annex I countries would be prepared to accept. The problem is how to devise an allocation formula which will be perceived as fair both by countries already developed (with a good position on the baselines) and those still developing (with urgent aspirations for future development). It may be easier to think of a fair process by which to arrive at an allocation than to agree in advance what the allocation should be.[32]

---

[31] EIA, *International Energy Outlook 2000*, table A.10, p. 179. Annex I countries' share of some other GHG emissions – for example methane – is lower.

[32] For an analysis of the fairness problem, and a fully worked out proposal, see Bartsch and Müller, *Fossil Fuels in a Changing Climate*, chs 13–14, pp. 226–79.

If Annex I countries cannot stabilize global GHG emissions at any level, the key policy question is not whether they should use 'flexible' or 'inflexible' tools, but whether the cost of even the most efficient tools available to them is justified by the likely benefits.[33] US Senate Resolution SR98 of 1997 demands that the United States not become a party to any agreement which mandates new commitments to limit or reduce GHGs for Annex I parties unless 'specific and scheduled' commitments are made for developing countries for the same commitment period. It is argued (by EU spokes-people) that Annex I commitments should come first because without them no approach to the developing countries is possible, and this was explicitly recognized in the UNFCCC. Developing country spokespeople argue that limits to their emissions are unacceptable because they would also limit their economic development. Designing emissions reduction targets for developing countries would be even more difficult than for Annex I countries. Prevailing 'conventional vision' estimates suggest that by 2020 China will account for over 40%, and India and South Korea another 14%, of all developing country CO_2 emissions.[34] Devising a single formula to fit these three countries is unlikely to be feasible, and it is arguable that the baseline and ration approach is fundamentally wrong for their situations. None is in equilibrium in relation to the others, in the internal structures of their economies, or in relation to their capacity to introduce and fund rapid technical change. Assumptions about the baseline for OECD countries in Kyoto opened up the problems of divergence described earlier for Annex I parties.

Looking for a lift

For Annex I countries, the Kyoto process is in trouble for three reasons:

- These countries have been slow to take action, so that even the modest commitments of Kyoto now look unachievable.

[33] See Nordhaus and Boyer, *Requiem for Kyoto*, pp. 122–6.

[34] EIA, *International Energy Outlook 2000*, table A.10, p. 179.

- The baseline structure has got out of step with reality: it neither provides a politically equitable sharing of adjustment costs nor reaches for economically efficient solutions. There are large practical and political problems associated with the flexibility mechanisms.
- There is still no real basis for serious negotiation with developing countries, without which the costs of Kyoto will be disproportionate to the likely benefits, especially for the United States.

It is not clear what will happen next. The Havana Charter of 1947 remained a façade for 50 years while the real business was done in GATT, which had been intended only as a subsidiary annex of the Charter. UNCLOS was not ratified until major elements dealing with seabed exploitation were substantially amended. The regime for reducing oil pollution at sea was developed incrementally over nearly half a century. The amount of time and effort that have been invested in the present text and structure of UNFCCC–Kyoto are such that it is unlikely to be renegotiated. The process could, perhaps, be 'helicoptered' to a new baseline year, leaving fewer contours to climb and a more even positioning for the individual climbers. The maps could be altered by changing the labels on the contours, so that the commitment period would move away from 2008–12. Finally, Kyoto could be left as a kind of 'North-West Passage' in the history of climate policy while the principal parties give priority to finding other routes to mitigate the risk of climate change.[35] These alternatives – 'fix', 'fudge' and 'façade'[36] – all face the same constraints: base years are arbitrary; dynamic adjustment of agreed targets is difficult to negotiate; the implementing policies can be very damaging if capital stock is destroyed; it may not be possible to trust all governments to recycle and redistribute revenue from new taxes. Fixing the present structure is probably impossible; fudging it merely postpones any solution. Maintaining the mission's façade while conspiring to look

[35] The North-West Passage was a supposed ice-free sea route from the North Atlantic to China and on object of many failed and disastrous exploration missions in the late eighteenth century. Its purpose was eventually better served by the building of the Panama Canal.

[36] See H. D. Jacoby '"Could Kyoto Work?', paper presented to conference on 'The Kyoto Protocol; The End of the Beginning', London, Royal Institute of International Affairs, 2000.

for other routes is the default mode for those seriously interested in developing real climate change policies.

What are the options for other approaches? Two are within reach of governments and could form part of a 'Kyoto-plus' strategy without necessarily being closely connected to Kyoto's quantified emissions assignments. Many studies have shown that existing taxes on energy consumption and energy-saving materials and equipment are not environmentally logical.[37] Typically, taxes on fuels differ according to the ability of their customers to pay (high taxes for motorists, low taxes for industry) or according to supply policy (typically anti-oil and pro-coal). Reforming these taxes to introduce a bias against carbon (without increasing the overall level of energy taxation) is intellectually not difficult. Even with no increase in the overall level of energy taxation, a carbon bias could induce switching from high-carbon fuels like coal to low-carbon fuels like gas or renewables.[38] It would, however, be bureaucratically difficult because tax systems are complex. It would also be politically difficult, because inevitably there will be winners and losers. These can best be managed within the framework of broader (and even more complex) tax reforms which are not lightly undertaken. The German Green party's efforts to promote ecological tax reform during 1998–9 are an example of the difficulties of advancing fundamental changes in tax philosophy. In the UK a 'climate levy' on fuel will take effect in 2000. It follows a reduction in 1997 in value added tax on fuels for household heating. The British government wanted to protect and improve the capacity of poor households to maintain adequate standards of comfort: (similar reductions on insulating materials eventually followed, but the underlying problem of poor-quality housing stock remains).

Most of all, energy tax reform would be politically difficult because to be effective it would require new taxes on coal in countries where coal matters. However, there is some scope for ingenuity and initiative: The European Commission's unsuccessful carbon tax proposals of

[37] See many studies by the Wüpperthal Institute, especially Ernst von Weizsacker and Jochen Jesinghaus, *Ecological Tax Reform: A Policy Proposal for Sustainable Development* (Atlantic Highlands, NJ: Zed Books, 1993).

[38] The OECD, in *Energy Prices, Taxes and CO_2*, suggests that, for the OECD, shifting existing taxes to a carbon basis would both reduce carbon emissions and increase GDP.

1993–4, and the existing carbon taxes in some EU countries, assume on good cost–benefit grounds that, since all carbon in the atmosphere causes equal damage, all carbon should be taxed equally, whatever fuel carries it. But other existing taxes on fuels are different proportions of their final price; there are different ratios of carbon to energy in different fuels, and even in competitive markets underlying fuel prices differ. The result is that a given tax per unit of carbon has different proportional effects (and therefore may be expected to have different effects on demand) for different fuels.[39] A tax which 'spreads the burden' of change equally among different fuels would take these factors into account. A compromise between equating the percentage impact on fuel demand and equating the taxes on carbon content would not be politically ridiculous.

A second way for governments to move would be to reduce energy subsidies and price controls that distort the environmental (as well as economic) options for consumers. This is particularly, but not exclusively, a problem in developing countries, as the IEA and others have shown.[40] Removing such explicit subsidies is not easy: no one likes to pay more for energy. Increases in transport fuel prices may bear heavily on the poorest section of the population, for whom transport (from outlying, low-cost suburbs) or fuel and light (in poorly insulated or cooled dwellings) takes a high proportion of their income.[41] The key point, as with taxes, is first to put a climate bias into the existing structure: reducing subsidies altogether (producing economic as well as climate benefits) can be a longer-run goal.

Neither tax reform nor subsidy reform needs to take identical forms in all countries. Teams taking this route to the sustainable plateau need not be closely tied together as long as they remain in contact. There might be competitive distortions (the famous 'leakage' caused by the relocation of

---

[39] This point is brilliantly illustrated in Bartsch and Müller, *Fossil Fuels in a Changing Climate*, fig. 11.6, p. 196.

[40] The IEA (*World Energy Outlook 1999*) presents case studies for eight non-Annex I countries. For these countries as a group, removing energy subsidies would reduce energy consumption by 13%; the study does not address the social or industrial impact on the beneficiaries of the subsidies.

[41] A quantified illustration of these effects is given in Bartsch and Müller, *Fossil Fuels in a Changing Climate*.

the supply of energy-intensive goods to non-Annex I countries) if differences among countries competing for international markets and investment were very large. There are some devices ('border adjustments' or contravening duties) which can be used to mitigate these. International financial agencies can provide technical help, general direction and incentives to other dealings with developing countries to bring them into the party. Such processes neither commit developing countries to constrain their growth nor pre-empt their ability to achieve 'cheap' GHG reductions in a world in which this will inevitably become a competitive advantage.

A third approach does not depend on generalized government actions, though it will work better if government constraints such as taxes and subsidies are set to facilitate rather than obstruct it. The United Nations Environment Programme (UNEP) and private sector organizations such as the World Business Council on Sustainable Development (WBCSD) are approaching industry sectors on a global basis.[42] The first step is a joint analysis of the sustainability challenges in each sector. Technological and policy opportunities for improving sustainability within the sector are identified, and the appropriate policy obstacles or incentives sketched out. These may include voluntary agreements or a combination of voluntary agreements and a package of relevant policy initiatives. How the policy initiatives are taken forward will depend on what forum is appropriate for the measure. Although studies in sectors such as transport, paper, and cement are emerging, it remains to be seen how coherent initiatives can be driven into the decision-making fora.

The sectoral approach has several potential strengths which are complementary to those of the international regulatory regime represented by UNFCCC, Kyoto and the analytical engines of the IPCC:

- Because it is led either by the private sector or by a UN international agency, it is non-national. The general international political considerations (US–EU, North–South) do not prevent the development of

[42] 'The WBCSD is a coalition of 120 international companies united by a shared commitment to the environment and to the principles of economic growth and sustainable development': *Signals for Change* (Geneva: WBCSD, n.d.), p. 4. Its general perspective is described in Stephan Schmidheiny, *Changing Course* (Cambridge, Mass.: MIT Press, 1992).

proposals. To the extent that a proposal requires intergovernmental agreement, the international political considerations will emerge and their cost to the progress of the sustainable development idea will be clearly identified.

- Because of the participation of private sector companies, some argument and challenge can take place about what is reasonable and practical and what the competitive effect of government action or inaction might be.
- The possibility can be created of *international* voluntary agreements. If conditions can be developed for voluntary cooperation by the leading international competitors in each segment of the sector, many of the competitive problems of 'leakage' and relocation which limit national approaches can be reduced. Chemical industries in one country will not need exemptions from costly environmental policies if their leading competitors are taking similar action. The problem is to find an international counterpart corresponding to the national environmental authority to accept and support the voluntary agreement by appropriate legal sanctions. This could take the form of MEAs, protocols under the UNFCC, or possibly even commodity-type agreements under the WTO.
- In the battle for corporate reputation, general objectives can be linked to practical actions within reach of the private sector, rather than being entirely dependent on the introduction of new regulation through international conventions.

Finally, there is scope for multilateral financial agencies (World Bank, Global Environmental Facility, European Bank for Reconstruction and Development) or national agencies (such as the National Economic Development Organization in Japan) selectively to target aid and finance to develop low- or zero-carbon energy supplies in developing countries. The technical and institutional lessons learnt from such projects can reduce their long-term cost and improve their long-term competitiveness with traditional forms of energy.[43]

[43] See Thomas Johansson and Susan McDade, 'Global Warming Post-Kyoto: Continuing Impasse or Prospects for Progress?' in *Energy and Development Report 1999* (Washington DC: World Bank, 1999).

Impact on oil

In Chapter 2 we described the important effect of income growth on energy demand, and the effect of relative fuel prices on the oil share of that demand. Modelling these effects for individual fuels requires heroic assumptions about policies which have not yet been decided: will climate mitigation policies be 'fair between fuels' in the sense that full recognition is given to their different carbon–energy ratios? Applying a uniform carbon tax to the carbon content of coal will have a larger effect on its price per unit of energy than for many oil fuels which are already taxed.[44] Under such a scenario, oil volumes in 2020 might not be very different from those predicted in the conventional vision; provided that the new taxes on fuels were efficiently recycled, there would not be too much effect on energy demand as a whole. However, if the new measures also include a general increase in the average level of tax (or restrictions of use) for carbon, then there would be an energy-saving effect which would reduce the demand for all fuels. Even so, the effect on oil revenue of the exporting countries could be within a margin of 10% of a reasonable 'conventional vision' since most of the volume effect of lower demand would fall on new and unconventional sources of oil, and on higher-priced gas, rather than on today's main exporters.

Summary: climate under the clouds

Climate change mitigation policies are necessarily international: to the extent that the planet warms up because of what we do, it does not matter where we do it. The developed countries cannot protect their climate without long-run support from the developing countries. As a model for their long-run support, the Kyoto Protocol has serious problems: it aims to assign national emissions rights which have to be translated into complex government intervention in the use of energy.

---

[44] The percentage increase for hard coal prices would be ten times that of gasoline in the *Primes* scenarios shown in the EU Shared Analysis Project: 'European Union Energy Outlook to 2020' in *Energy in Europe* (Brussels: European Commission, November 1999), table 4.2, p. 73.

Different countries may adopt different policies and measures to induce their citizens and businesses to fulfil the national objectives. It is not yet clear what the main methods used will be, or how different national measures will relate to one another in practical detail. Meanwhile, reality has diverged so far from the 1990 baseline as to make the commitments for 2008–12 much more difficult to achieve for some key nations than for others. Flexibility mechanisms (all involving the creation of new institutions, processes and bureaucracies), may reduce the costs of compliance but cannot cancel them out. Although nothing is certain, the ' balance of probability' is that the Kyoto commitments to reduce emissions will not be fulfilled as written. The impact on the 'conventional vision' for oil will be within the range of normal forecasting uncertainty for the period 2008–12.

For the longer term, the picture is entirely different. The difficulties of implementing Kyoto increase the longer-term uncertainty for oil and its users. The delays which have occurred in making restrictions of GHG emissions effective will increase the severity of the measures which will eventually be needed to limit GHG concentrations in the atmosphere to any given level. Operations during 2000–20 may track close to the 'conventional vision', but plans and investments made during this time will more and more need to follow the trend towards a different future. It is certain that the efforts to develop effective climate change mitigation policies will continue. They will be driven by governments practising responsible prudence, and by publics steadily made more aware of the costs of adapting to a climate increasingly likely to change in uncertain ways. The growing scientific evidence for climate change will be connected with public attribution of 'extreme weather events' such as hurricanes and floods to climate problems. More and more influential actors will share the perception that 'something needs to be done' – by businesses as well as governments and citizens – even if agreement on action is difficult to reach. More and more enterprises in the private sector will contribute to the development of cost-effective policies by participating in the process and promoting global strategies based on technical innovation and business interests.

The social acceptability of oil

Health, safety, pollution and climate issues are not the only factors which matter for the acceptability of oil. The previous section described the development of major agreements among governments to establish common and cooperative action against cross-boundary threats which, through phenomena such as the ozone layer and the global temperature, could affect the health, lifestyles and economies of individual countries. In taking these actions governments were influenced by public opinion as well as by scientific advice and moral self-interest. Media reporting of oil spills, and public anger at their consequences, contributed to governments' willingness to agree on the MARPOL convention on marine pollution and the environmental provisions of the Law of the Sea, UNCLOS. Expressions of alarm by NGOs over threats of climate change contributed very specifically to governments' willingness to discuss the UNFCCC at Rio. Once at Rio, governments were subjected to remarkably efficient lobbying by the NGOs concerned: this undoubtedly influenced the form and detail of the outcome.[45] The next section shows how similar mechanisms are developing to challenge the acceptability of the supply and use of oil on grounds of the effect on sustainable development, social responsibility of corporations, and the human rights of those affected by corporate or government activities in the oil sector.

Sustainable development

The idea of sustainable development is, in a sense, a revolutionary synthesis:

> Sustainable development is development that meets the needs of the present without compromising the ability of future generations to meet their own needs. It contains within it two key concepts:
>
> - the concept of 'needs', in particular the essential needs of the world's poor, to which overriding priority should be given; and

[45] See Brenton, *The Greening of Machiavelli*, pp. 256–7.

- the idea of limitations imposed by the state of technology and social organization on the environment's ability to meet present and future needs.[46]

The UN Conference on the Human Environment in Stockholm in 1972, the Earth Summit at Rio de Janeiro in 1992 and the UN Conference on Social Development in Copenhagen in 1995 expanded this idea in many directions. An extensive literature has developed to address the problems of agreeing on what is equitable between generations and what is an acceptable degree of transformation in the natural environment to meet the 'need' of everyone – but especially the relatively poor – for a better life.

For people in business, the idea of sustainability is in the economic sense entirely familiar. Outside the centrally planned economies and some government accounting systems,[47] it is normal to account for the depletion of capital and to make provision for replacing it. There are complicated controversies about how to measure the capital in finance and business – whether at its original cost, at the cost of physically replacing it exactly, or at the cost of more modern plant which would produce the same result in a different way. The same issues arise regarding the sustainability of the natural environment: how much substitution should be accepted as consistent with 'sustainability'?

Economists would like to express 'acceptability' through the pricing mechanism, but obvious problems arise because, unlike the situation of a firm, the situation of a society with regard to its physical environment is not easily measured by prices: the environment is sometimes indivisible; its attributes cannot be bought and sold. Environmental damage (like some damage to human life and capacity) is irreversible. If there are markets, the markets are incomplete and the prices omit real 'external' costs.

There are other useful analogies for business. Businesses are concerned with their reputation because they expect to repeat their transactions with each other and with investors, customers and suppliers. Business managers need to be trusted. They also need the confidence

---

[46] World Commission on Environment and Development (the Brundtland Commission), *Our Common Future* (Oxford: Oxford University Press, 1987).

[47] Such as the UK's until 2002.

Table 8.3: ' Genuine savings', excluding CO_2 effects, 1997

Country	% of GDP
Algeria	29
Angola	3
Indonesia	9
Kuwait	–24
Nigeria	–10.5
Saudi Arabia	–13
Venezuela	10
Iran, Iraq, UAE	n/a
For comparison:	
East Asia and Pacific	*31*
France	*12*
Germany	*14*
USA	*10*
UK	*8.5*

Source: World Bank, *World Development Indicators 1999*, table 3.15.

and loyalty of their staff – which are likely to be weakened if too great a gap arises between what the business is trying to do and what its members are prepared to accept as citizens. In the context of international relations, these difficulties have been classically approached by negotiating agreements to protect common interests – fish stock, biodiversity, and (as discussed above) the climate.

The division between Annex I countries and the rest in the climate change process illustrates a common North–South dilemma: the developed countries place the greatest burden on the global environment, and have the greatest technical and financial capacity to substitute for it or to deal with any long-term damage. The poor countries depend in many cases on depleting natural resources – including common global resources – because they are in the early stages of building up the technical and human capacity to improve their lives by other means. The main oil-exporting countries fall into the latter category. The World Bank has tried to estimate 'genuine domestic savings': the sum of the normal measure of domestic savings in the economy, plus expenditure on education (forming human capital), less energy and mineral resource depletion and less damage to the global climate through CO_2

emission. Even excluding the last factor, many oil exporters did not look economically sustainable in 1997, the last year estimated (see Table 8.3). Countries with negative savings are by definition not able to sustain their current level of activity.

These measurements are experimental, and for one year only, but they illustrate a particular problem of the oil-exporting countries. The various attempts by the World Bank and others to estimate indicators of well-being other than GDP continue – and they often indicate that countries with low GDP per capita are not quite so badly off relative to the high-income countries as GDP comparisons would suggest.

'Sustainable development' has put 'nature' alongside ' humanity' as something to be cared for in the process of development and by international actions to be supported by government agreements and internationally accepted behaviour. But that is not the only trend affecting the 'acceptability' of oil.

Idées sans frontières

The last decade of the twentieth century saw dramatic changes in the cost and availability of communication. Technology reduced (and is still reducing) the cost of transmission and equipment. Competition and privatization policies are enabling improvements in efficiency and attracting new investment in many countries (including many developing countries). Networks are expanding; ownership of, or access to, communication is increasing. Between 1990 and 2000 the number of main telephone lines increased by nearly 90% worldwide; international telephone traffic trebled; there are now two-thirds as many mobile telephones as main lines, and nearly 400 million Internet users.[48] The development of wireless access for cellular telephones will expand the freedom of individuals and organizations to communicate across frontiers without the use of networks.

The political effect of this expansion is yet to be seen. The 'fax factor' in the Iranian Revolution of 1979 was limited by the capacity of the network, the cost of access to it, and the need for fax senders to

[48] ITU Telecomm Indicators, <www.itu.org/ KeyTelecom99.htm>, 7 April 2000.

identify and locate their targets. The NGOs at the Rio Summit in 1992 used the Internet and email effectively to out-communicate some national delegations in dealing with their home governments. More recently, the Internet has been the principal tool for NGOs to advertise situations such as the human rights violations associated with the Yadana pipeline route in Burma, the problems of Ogoniland, and the displacement of people from large power or infrastructure projects anywhere. Advertisements solicit responses in the form of letters to political representatives, consumer boycotts and demonstrations. Small groups can make big noises about bad events.

Private international relations

The explosion of communicating power, the activism of NGOs, and the accountability of private sector managements to shareholders sensitive to ethical issues create a new arena of international relations. Political objectives can be achieved over borders, and values transmitted or upheld in any part of the world without necessarily passing through the interstate process of government negotiations, conventions, and the operation of international governmental institutions. Obviously these are not superseded; NGOs and concerned or affected groups will still seek remedies or progress through the classic governmental routes. But these are slow, complex and subject to the balancing of bargains across a wide range of policy and institutional agendas. For single issues, direct action in the private sector can deliver something visible, sometimes quickly. For actors, such as oil companies challenged by new issues or new advocates, failure to respond in the private arena may lead to failure to influence the outcome in the official arena. Resolving uncertainties by private sector initiatives and dialogue may lead to better outcomes for those concerned than relying entirely on government negotiators in the state-to-state processes. Figure 8.1 illustrates the idea of a broad system with both government and private components.

Figure 8.1: Public and private international relations

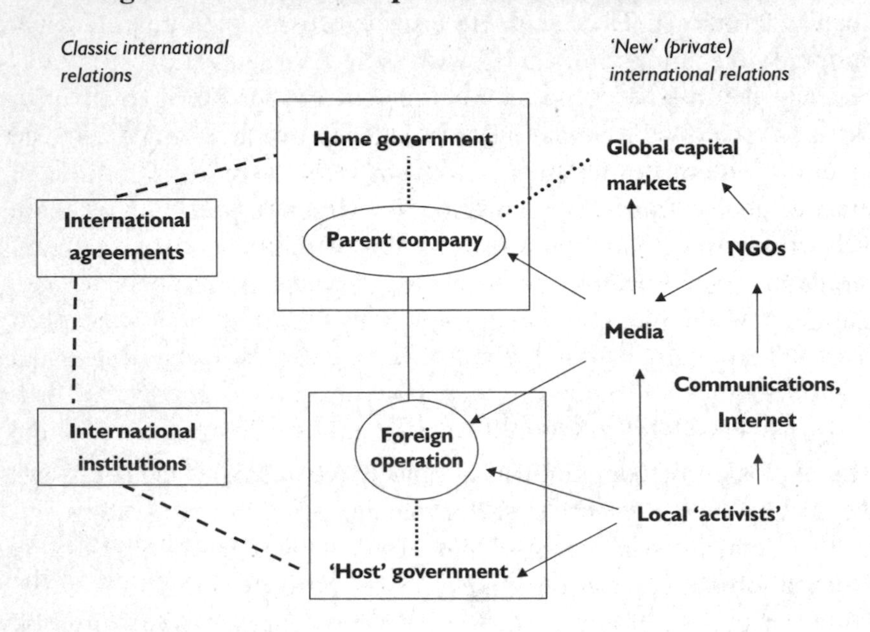

Responsibility for oil supply in global civil society

In the free-for-all of private international relations, the proper responsibilities of private sector corporations are confused and controversial. What is in question is not their technical or economic responsibility but their social responsibility. Moreover, it is responsibility exercised in an international 'civil society' where values and interests are in conflict and where intergovernmental agreements do not cover all the cross-border contingencies which face transnational companies.

Multilateral agencies have a slightly simpler task than companies, since their social and environmental roles may be prescribed by their charters, but there is generally room for controversy over interpretation. The World Bank, for example, has faced criticism over the environmental impact of projects undertaken in the 1970s, and over the social and human rights impacts of controversial hydroelectric schemes. NGOs have tried to use their status at the UN to become involved in Bank appraisals, as well as advising it, on request, on general issues and lobbying it on particular ones. The Chinese government in July 2000 withdrew an application for a $40m loan from the World Bank to resettle Chinese farmers in Tibet in the face of objections from a pro-Tibetan NGO.[49]

Companies can face criticism because of actions taken by governments to provide infrastructure for projects in which they are engaged (as in the Yadana gas pipeline in Burma), or because of the actions of government security forces protecting the company's operations, as in BP's operations in Colombia.[50] Objections to Chinese 'encroachment' on Tibetan society led to demonstrations against BP Amoco, not because of any activities of the company in Tibet but because it announced plans to take a small shareholding in a Chinese company which had such operations. Talisman, a Canadian company undertaking oil production development in the Sudan, was heavily criticized because the taxes and royalties it paid from the project revenue to the Sudanese government were allegedly used to finance an oppressive civil war. Similar criticisms were made in Nigeria during the civil war there in the 1960s and in the Angolan civil war today. They have parallels in the criticism made of foreign companies investing in South Africa during the apartheid regime.

In some of these cases the criticism of the company is initiated by local activist groups or by émigrés. The 'Free Burma' Campaign shows that such groups can achieve good results, especially when the foreign operation concerned (as in Burma) is new or small relative to the company's interest in the United States or Europe where such campaigns take effect. Texaco, Pepsi-Cola and Carlsberg withdrew from Burma following campaigns targeted against them on the grounds of the circumstances surrounding their specific operations. Some companies, and some NGOs, seek to deal with such criticisms by strict codes of conduct for their own operations, and use the buying and taxpaying leverage to induce 'good behaviour' in contractors and in the government agencies with which they deal. This does not answer

[49] See 'China Drops Plea to World Bank to Fund Tibetan Scheme', *Financial Times*, 8–9 July 2000, p. 1.

[50] For a description of site-related problems, see John Mitchell, 'Human Rights: One More Challenge for the Petroleum Industry', in John Mitchell, ed., *Companies in a World of Conflict* (London: Royal Institute of International Affairs /Earthscan, 1998).

the wider question of 'guilt by association' with a regime which is regarded in North America and Europe as oppressive, violating human rights, or failing in its duty to alleviate poverty and conduct open government by standards of honesty and fairness which would be acceptable, or at least not unprecedented, in developed countries. These issues go beyond the ideas of protecting the global environment, contributing to sustainable development, or even protecting human rights. It looks as though the foreign companies are being used as missionaries to carry the values of the investing countries into the 'host' countries. The companies are not allowed to plead moral relativism in politics any more than they are allowed to practise double environmental standards for the rich and the poor.

NGO who?

The UN Statute for NGOs establishes criteria for recognizing NGOs and giving them access to certain UN information and meetings.[51] Recognized NGOs must be non-profit-making, not support violence, not be established by governments, not campaign against UN programmes, have an identifiable organization (preferably democratic), and not be a political party. Some NGOs, such as trade union or trade and industry federations, promote the economic interests of their members; some focus on single issues; others – the great transnational NGOs – deal with a wide range of issues.[52] Not all NGOs are recognized by the UN. Some transnational organizations, such as terrorist groups or criminal alliances, obviously act outside both international and national law.

In discussing the 'acceptability' of oil – or anything else – the NGOs which matter are those that are advocates for policies which they claim are in the general interest of humankind or of 'global civil society'. They seek legitimacy for their specific causes by reference to

[51] Economic and Social Council (ECOSOC) Resolution 288 X B, 1296(XLIV)1968 and 1996(310).

[52] For a description and discussion of definitions and classes of NGOs, see Peter Willetts, 'Political Globalization and the Impact of NGOs', in Mitchell, ed., *Companies in a World of Conflict*.

this wider interest and the values which underlie it. There is obviously room for argument, and values change. In democratic societies, core values include due process, fairness, exposure of argument to challenge and open debate, and acceptance of formal social processes for resolving conflicts – such as courts, fines and prisons to enforce law and contracts. Questions of who elects NGO spokespeople, how they are funded and to whom they are accountable are in some senses details so long as they accept the formal processes of society – including the society of states – as the means of resolving disputes. NGOs that promote lawbreaking for demonstration purposes but accept its consequence for the lawbreakers are exercising an extreme form of protest but do not challenge society's right to enforce the law. What matters is not so much how they elect their representatives but whether their cause is arguably consistent with the core values of society and how much deep offence or damage is caused by their methods. At the other end of the spectrum are terrorist and guerrilla groups who do not seek legitimacy from existing society. They break laws and try to avoid the consequences. In between there are problematic situations of groups who promote values which are widely shared outside the country but are suppressed – sometimes illegally – in the countries where operations occur. Here real conflicts of values between societies arise. Because transnational companies link societies with different values, their activities become a bridge on which battles about values are fought, whether the values are environmental, social or political.[53]

Ethics: the investors' last stand

The liberalization of international capital movements since the 1980s in most OECD and many developing countries has opened a new international dimension to management responsibility to shareholders, because those shareholders may now be anywhere. European, Russian and even Chinese enterprises seek to list their equity on the US

[53] For a review see Peter Willetts, 'Transnational Actors and International Organizations in Global Politics', in John Maylis and Steve Smith, eds, *The Globalization of World Politics*, (Oxford: Oxford University Press, 1997).

financial markets, in London, Frankfurt, Tokyo or Paris, and to issue publicly quoted bonds outside their own countries. This requires, in many cases, more disclosure about their company's activities than was necessary when shareholders were domestic investors alone. It requires an accounting by management to financial analysts in the major market. Behind the analysts stand not only the individual investors and 'retail' investment funds but pension funds and insurance companies responsible for investing billions of dollars of individual savings.

Critics of 'short-term profit' motives often overlook the fact that, for these investors, profits are measured not only by the flow of dividends, but by movements in share prices – especially their movement relative to the stock market as a whole. Stock prices reflect the stock market's understanding of the company's potential for long-term growth and the risks it faces, as well as expectations of the next quarter's dividend. Investment managers like to know whether the company is following and prudently anticipating changes in its operating conditions, what competition it faces, and the likelihood that it may face government or public action somewhere in the world which may damage its stock price. This stock market exposure provides a 'weak' reason for management to attend to environmental and social matters and an incentive to report on environmental and social impacts and performance. These reports, and the interests of analyst and fund managers, are tending to change the onus of proof on the manager. Traditionally, management had to defend 'good citizenship' as being in the long-term interest of the shareholders, according to management judgement. With environmental and social reporting and attention to it by investors, management has to respond to external judgements about what good corporate citizenship requires.[54] NGOs play a part in this through canvassing analysts and fund managers and through buying token shareholdings in companies. Such token holdings give them, under most North American and European company law, a right to attend annual meetings and ask questions of the management. Subject to some limits, they may also have a right to move resolutions.

[54] For a brief overview see Halina Ward, *Corporate Citizenship: International Perspectives on the Emerging Agenda*, conference report, June 2000, London: Royal Institute of International Affairs.

Pension funds generally have fiduciary obligations to maximize the sustainable value of their investments, and may be influenced by ethical factors only if company behaviour puts its share price or dividend at risk – a 'weak' argument for attention to acceptable company behaviour. Nevertheless, pension funds may have considerable discretion. Where pension trustees are elected by fund members (as is the case with many public employee or university pension funds in the United States), these members may require attention to ethical questions. US pension funds' refusal to hold stocks in companies investing in South Africa were among the 'private' sanctions applied to South Africa in the apartheid era. UK regulations now require British pension funds to declare the extent to which such questions are taken into account in selecting stocks.

Some retail investment funds deliberately set out to avoid investment in companies involved in 'unethical' activities – such as arms or tobacco manufacture – and companies engaged in inevitably damaging environmental practices (such as destructive forestry). The record of these ethical funds is sufficiently good for them to argue that those who invest through them are not worse off and are perhaps better off. There is some argument that companies that take care of the details of their environmental and social impact also take care of efficient operations and good relations with customers: the 'strong' argument for ethical investment.[55]

For the oil industry, a defining moment in shareholder activism was the BP Amoco annual meeting of 13 April 2000. A group of Greenpeace supporters, under the name SANE BP, and Trillium Asset Management Corporation, an 'activist' US ethical fund, moved a resolution to halt BP exploration and development in the Arctic, avoid work in the Alaska National Wildlife Reserve (should it ever be permitted) and increase investment in renewable energy sources. The resolution was rejected, but the sponsors secured support from 13.5% of the votes cast (about 8% of the shares issued). Well-televised Greenpeace demonstrations on the North Slope of Alaska on the previous day ensured that the

[55] See 'Companies Come Under Pressure to Alter Course', *Financial Times* 8–9 July 2000, p. 8 (UK edn); 'Morality Pays', *The Economist*, 8 July 2000, pp. 122–3.

resolution received wide publicity. The votes cast, representing about £8 billion of investment, were clearly not just the votes of token shareholders.[56] Private sector corporations with large US and European shareholdings can expect controversial projects and behaviour to be challenged in the future.

The next oil crisis: behaviour

As private sector trade and investment grow, and as private international relations become more visible, all international trade and investment activities will continue to be affected. What makes the trend particularly important for the oil and gas sector is the location of so many new exploration, development and infrastructure projects in countries where the 'value gap' between them and North America, Europe and Japan is likely to be large and long-lasting. The oil and gas-rich countries of the Transcaucasus, Central Asia and West Africa (as well as Burma and Indonesia) are all subject to criticism for human rights violations by their governments, as documented by the annual reports of the US State Department, Human Rights Watch and Amnesty International. Many of these countries also score poorly on privately published indices of 'good governance'. Issues concerned with the land rights of indigenous peoples and their political ability to protect such rights affect oil exploration and development in Colombia, Ecuador and Brazil. Offshore operations (as in Angola) may keep companies out of some local security problems but do not distance them from questions of how the government spends the money and how contracts to supply goods and services to development projects are allocated. Investors in established oil-producing countries such as Indonesia, Angola and Algeria can expect similar challenges. Criticism can be based not only on the UN human rights conventions but also on the Geneva conventions for the treatment of civilians in civil wars. The oil companies' dream of being readmitted to the major Middle East producers, if it were ever realized, would be disturbed by the same kinds of issue: in

[56] Details of the resolution, the directors' counterstatement, and the polling were posted on the BP website, 'shareholders' information'.

Iraq, unimaginable changes would have to occur in the regime; even Saudi Arabia is not party to the UN Convention on Human Rights and stands at a tangent to much of the developed world in its lack of democratic institutions, its treatment of women and its use of severe physical punishments including the death penalty.[57] Altogether, about 40% of the world's oil production comes from countries where human rights, as defined by the UN conventions, are not recognized because the country has not acceded to the conventions, or from countries which have acceded but where human rights nevertheless appear to be seriously and frequently violated. The corresponding proportion for new exploration and development opportunities for the private sector would probably be higher.

Similar problems affect many of the large international gas pipeline projects which will be necessary if the 'conventional vision' of increasing gas use in developing countries is to be achieved: The Yadana line for Burma's gas exports is an established case. The proposed Baku–Ceyhan oil pipeline through Turkey has raised questions about Turkish treatment of its Kurdish citizens in the areas through which the line will pass. The proposal for a gas pipeline to export Turkmen gas to Pakistan through Afghanistan has been abandoned: Afghanistan under Taliban rule is regarded as a 'rogue state'. The pipelines proposed for the import of gas to China from various parts of Russia could be a focus for protest if, as in the Tibetan case, they become associated with internal Chinese policies which are disliked by groups outside China concerned with human rights in that country. All these instances add up to a wide panorama of 'acceptability' issues affecting the supply of oil over the next 20 years.

[57] The United States is also criticized by human rights groups for maintaining the death penalty.

Chapter 9

Challenges and choices

The 'new economy' of oil is a mixture of continuity and change. Some of the structural changes are the result of accumulations of past trends to the point where the structure itself is altered: the crumbling stone walls eventually open a new horizon. Others are the result of new forces outside the oil sector. The Internet and communications revolutions are transforming political life and especially the international exchange of opinions and values. These set the context for public policy and the strategies of private corporations. Questions of the acceptability of oil are becoming as important as worries about its availability – at least for the next 20 years.[1] Analysis of supply needs to be matched by understanding of demands, not just for fuels but for environmental and social results. The 'new economy of oil' is a *political* economy, not just in the old geopolitical sense of calculations of national interest, or the newer aspirations for sustainable development, but also in terms of the expansion of the real freedoms that people enjoy.[2]

The 'conventional vision' of oil and gas demand, supply and prices through to 2020 described in Chapter 2 is a credible projection of 'business as usual' and a basis for policy analysis. Energy demand will grow more slowly than the world economy. Supplies will grow to match, with a shift towards gas in most regions and markets. The transport sector will still call for less than 60% of the total projected oil supply by 2020. The balance will have to compete with coal, gas and non-fossil fuels. Constant or falling coal prices will anchor prices for fuels for power near to their 1986–99 averages. Oil and gas exporters will continue to compete to bring their reserves to market. Most of the

[1] For a longer-term perspective in which availability may re-emerge as an issue, see Pierre-René Bauquis, 'Un point de vue sur les besoins et les approvisionnements en énergie à l'horizon 2050', *Revue de l'Énergie*, no. 509 (1999).

[2] See Amartya Sen, *Development as Freedom* (Oxford: Oxford University Press, 1999), introduction, pp. 1–11.

growth in demand for oil will take place in developing countries which will become more important than developed countries as a market for Middle East oil. Security against disruptions will remain a problem for which most Asian importers are currently unprepared. Fossil fuels will increase their share of energy supply (to nearly 90%) as nuclear power stands still or is phased out in the major countries now using it. Greenhouse gas emissions will rise inexorably in these projections, which do not allow for severe policies to mitigate their growth. Variations in assumptions about rates of economic growth and the trend in energy efficiency would shift the reference projections slightly.

The reference cases look like useful tramlines,[3] projections carrying all the forces of continuity of supply and demand from the past to the future. Below the track, however, there needs to be some heroic engineering of new foundations to carry the route over uncertain ground of availability of particular fuels in particular places – especially natural gas for Asia. Above the track, the landscape is changing and crosswinds are beginning to blow. Travellers and the policies that carry them forwards have some new questions to answer as they move forward in search of continuing acceptability.

Questions of availability

Oil production capacity: cyclical risks

The year 2000 saw the end of the massive surplus of oil production capacity which had dominated the world oil market since 1980. The price fluctuations of 1998–2000 reflected relatively small margins of surplus and 'short' capacity, magnified by pro-cyclical production policies by OPEC member governments. For the future, the oil-exporting governments are in a 'new economy' in which their decisions about expanding capacity are as important to their future oil revenues as their ability to collaborate during temporary surpluses.

We argued in Chapter 3 that, through to 2020, oil supply is unlikely to be constrained by lack of oil reserves or rapidly rising costs of production. Differences in revenue needs, production capacity and

[3] US English: tracks for streetcars.

oil reserves among the principal producers will make it difficult for the key oil-exporting governments to agree on an orderly sharing of capacity expansion. They are bound to compete for revenue by expanding capacity in the long term. Chapter 6 argues that this competition will prevent the emergence of a Middle East oil cartel capable of dramatically increasing oil prices above their 1986–99 average of slightly above $20 (1999$) per barrel.

It would be surprising if the oil exporters correctly guessed the medium-term demand for oil and the development plans of other suppliers, so that 'investment cycles' are likely. A high proportion of the potential for new oil production capacity depends on the investment decisions of governments and state enterprises.

Table 9.1 shows the principal oil production capacity expansions projected in the EIA (*International Energy Outlook 2000*) reference case. All depend on government decisions, though in some cases governments may seek investments by the private sector to implement them.[4]

Whatever the long-term forces affecting competition to expand production, it is clear that the path to 2020 will be affected by the future of US/UN sanctions on Iraq and other oil exporters and by the development policy of the Saudi government. It is quite possible (and the 'conventional vision' assumes) that both countries will expand capacity, competing with each other for market share. It is possible that while Iraqi expansion is blocked, Saudi expansion may be delayed. There is another reason to delay expansion: against expectations, most or some Annex I countries may adopt serious climate change policies which (for some of the reasons described in Chapter 8) may bear more

[4] At the time of writing in 2000, Saudi Arabia and Mexico showed no sign of relaxing their upstream state monopolies on oil production. Contracts for oil development projects for the Iranian National Oil Company provide the foreign partner with an opportunity to earn fees rather than equity returns, in projects decided by the state company. Ten years of talk have not yet produced large investment opportunities for private sector companies in Kuwait. In Iraq, private sector investment is the key to planned expansion but is frustrated by sanctions. For Venezuela, the expansion figures include extra heavy oil developed by private investment, but the programme of opening up conventional oil exploration and development opportunities to private companies under production-sharing terms seems, in 2000, to have come to a halt.

Table 9.1: Projected expansion of oil production capacity, 1998–2020

		Capacity, mb/d	Increase, %
State monopolies	Saudi Arabia	10.7	28
	Mexico	0.4	1
	Kuwait	2.6	7
	Iran	1.6	4
Total above		*15.3*	*40*
Other OPEC	Iraq	3.4	9
	Venezuela	1.4	4
	Other	4.9	13
Total above		*25*	*66*
Total world		38	100

heavily on oil demand than on demand for domestically produced fossil fuels such as coal. It would be rational for oil exporters to delay new investment until these policies become clearer. A coincidence of 'wait-and-see' investment policies among governments of oil-exporting countries could lead to a two–three-year scenario of a close balance between oil demand and production capacity. 'Wait and see' would be an easy option with medium-term prices in the $25–30 range – above the 1986–99 average and above the conventional vision for this period. Such a scenario would provide better economics for natural gas development and infrastructure and expanding non-OPEC oil production, and would also encourage the introduction of more efficient technologies and new fuels in consumer markets. The demand for oil in the latter part of the period (2010–20) would be reduced accordingly. The 1999 conventional vision of weak oil prices at the beginning, recovering to the 1986–99 average or above by 2020, would be reversed. Exporters with high oil reserves would face a combination of lower prices and delayed production.

The drive of Asia

Projecting into the twenty-first century involves projecting out of the familiar era when the majority of energy consumption took place in the OECD countries. In future, the uncertainties of energy demand outside the OECD will be more important than those within. The collapse in oil demand during the 'Asian' financial crisis of 1997–8 was a foretaste of future surprises. Some non-Asians may have been surprised by the speed and degree of recovery in the Asian economies.

Asia is not the only source of uncertainty. The different 'conventional vision' projections of energy demand analysed in Chapter 2 showed important differences between the United States and Europe. Some of these reflected uncertainty about whether the productivity gains of the 'new economy' in the United States will be replicated in Europe, the rapidly developing Asian economies and some Latin American countries. This is not primarily an energy question. The low costs and competitive markets for energy in the United States in the 1990s may have contributed to its recent exceptional economic growth, but a variety of other factors are obviously important. One key question is whether the 'new paradigm' in the United States depends on a combination of technology (which can be transferred) with social and institutional factors which are more difficult to replicate and may be welcomed only reluctantly in other countries. The effect on energy demand is difficult to predict. The more certain result would be geopolitical. If the 'new paradigm' is added to population growth and high savings and investment rates in developing Asia, the balance of power will shift from its present fulcrum. Among other things, the ability of the US administration or Congress to lead partisan sanctions against oil-exporting countries would diminish.

Fuel for transport

Chapter 4 describes the beginnings of a revolution in the transport sector in general and the vehicle industry in particular. For the first time in almost a century, competition is beginning to develop among vehicle technologies which will require different fuels to be delivered differently. Most of the possibilities will still require petroleum fuels or hydrogen, for which petroleum remains a competitive source. This competition is not the result of any imminent shortage of petroleum supplies for the transport market. It is a response to new opportunities arising from technical developments and to new demands for cleaner fuel, quieter vehicles and less congested cities.

Demand for electricity and choice of fuel for power

The composition of energy demand in the rapidly developing countries is more uncertain than that in developed countries. In the 'conventional vision' projections for developing countries, the projected rates of economic growth drive the projections of growth in energy demand but also assume that electric power supply will grow at rates of 4–6% to support the economic growth rates. This growth of power supply in rapidly growing economies depends on investment in expanding the power sector and the fuel infrastructure which supplies it. In many developing countries the pricing structure for electricity has been distorted by controls and subsidies, which exaggerate demand and also frustrate supply. The power industry receives neither the margins to finance investment in capacity nor the incentives to choose the potentially cheapest plant. For most developing countries, electricity *price* reform (withdrawing subsidies) is necessary to support continuing high rates of economic growth. To the extent that the reforms increase the price of electricity, it may be used more efficiently (depressing past trends). The question of electricity *structural* reform is more complex.

The US and UK models of regulating the electricity industry promote competition among independent producers and offer consumers a choice of supplier by providing 'open access' under cost-of service tariffs to national grids. The effect has been to remove the protection which previous regulatory regimes gave to investment in power production, either through price regulation (in the United States) or through state ownership (in the UK). Investment has tended to fall because investment in 'spare' capacity is no longer protected. Diversification of supply and the management of risks are decentralized to the market, with continuous adjustment through short-term trading. Similar changes are involved in electricity reform in Japan and in the EU electricity directive of 1996.[5] In each case the changes have taken a somewhat different form, depending on each country's different starting point.

[5] In Japan, the Electricity Business Act 1995, leading in 1999–2000 to industrial and large retail markets being opened to competition; in Europe, by opening 23% of each domestic market to competition by 1999 and one-third by 2006. France has not, at the time of writing, implemented the first phase of the directive.

What is common to all is the existence of a mature industry, in some cases with large regional or national grids and initial high surplus 'reserve' generating capacity, and rates of growth in demand around 1.5% a year.

For developing countries in Asia, growth in demand for power is projected in the 4–5% range. Many countries start with shortages, not surpluses, of generating and distribution capacity. The challenge is different. Reforms copied from the United States or UK, which involve privatization, bundled with liberalization and the promotion of competition, may make large-scale new projects more difficult to finance because traditional throughput and price guarantees cannot be found in the short-term competitive markets which the structural reform will create. The problem spills over into infrastructure investments for rapidly growing fuel imports. Finance for such projects has traditionally depended on take-or-pay contracts guaranteed at the importing end by a dominant marketer. In the main Asian developing countries the growth in demand for fuel for power outstrips local availability of supply, and new infrastructure investment will be needed for imports – of, for example, LNG and pipeline gas to East Asia. These have traditionally been secured by long-term contracts with fixed or indexed prices. The 'conventional vision' projection of power supply and the development of gas imports for power in Asia presume that power and infrastructure investment will be made in a timely fashion. This would not be guaranteed automatically by application of the US and UK models of structural reform in the electricity sector.

The speed and effect of structural changes in the electricity sector in rapidly developing Asian countries are uncertain. But the numbers involved are large and will increasingly affect the demand for globally available fuels like oil and coal. Power generation in the 'conventional vision' grows from around one-quarter of global gas demand in 1999 to over 40% in 2020.

Natural gas imports

Around 40% of the projected increase in natural gas use[6] – about 35trn cu ft in 2020 – is projected to take place in developing countries and to

[6] These fractions refer to the reference case in EIA, *International Energy Outlook 2000*.

be supplied by imports. This involves expanding existing or new import infrastructure in countries like Korea and Brazil, and establishing new infrastructure in countries like China and India. Long-distance gas transport requires large infrastructure investments, either in pipelines or in LNG terminals and vessels.

At the supply end, there is potential competition among export projects in those countries where known gas reserves exceed what is needed to support current production: Algeria, Australasia, Azerbaijan, Russia (in various regions), Indonesia, Iran, Kazakhstan, Nigeria, Norway, Trinidad, Turkmenistan and Venezuela.

Unlike national gas markets, the international market has no potential regulator to impose either competition or a central plan on international trade or investment. The Energy Charter Treaty and the related model agreement for gas transit provide some principles and language for international agreements covering pipelines, but these do not apply to LNG investments (and few Asian exporting or importing countries have signed the treaty). There is no international price for traded gas; regional markets have developed different pricing methodologies – a commodity market in the United States and a long-term, oil contract market in continental Europe.

Unlike the international oil market, the infrastructure for long-distance gas imports requires large initial investments. The economics are subject to strong economies of scale at higher levels than for oil. Transport costs account for a higher proportion of the final price. The traditional basis for the financing of LNG import infrastructure has been long-term take-or-pay imports into markets regulated to enable importers to recover their costs. The introduction of competition into the electricity and gas sectors (described above) could make it difficult to write such contracts in the future in many countries.

Without a coherent model for international gas trade, regional business models are developing in an ad hoc fashion. Chapter 5 gave some examples from developing countries. What happens in the major potential markets of China and India remains to be decided. Such business models are difficult to define when they are in conflict, or when the underlying internal markets follow different paths, some orientated towards competition and some towards protecting investment.

Policies and structural measures alone will not enable developing countries to increase the share of gas in their energy markets through imports. The price of gas must compete with the price of local or imported coal, and with the increasing surpluses of residue from an oil-refining industry tilting more heavily to the manufacture of gasoline. The cost of gas transportation and distribution needs to be further reduced by technical developments, and the owners of gas reserves need to accept prices which allow gas to compete with other fuels. The dilemma is that the competition which encourages cost reduction also adds to the risk of long-term infrastructure investments.

New questions of acceptability

In contrast to the new foundations of availability discussed above, the new questions of acceptability arise from political and social choices made outside the energy and oil sectors. The energy security calculus has been transformed by geopolitical changes. For the United States and its allies, international trade and investment in energy has become a foreign policy opportunity rather than a security threat. Environmental changes have become international through the emergence of the threat of climate change. This will change the direction of the demand side of the tramline of conventional vision. The social and political impacts of major project developments in small or weak countries call forth responses on issues of human rights, good government and democracy from customers, investors, NGOs and governments where they are subject to the pressures of democratic and media-dependent public opinion. These responses fall directly on to the international actions of companies from the international sharing of news and opinion through global media and communications, catalysed by NGOs and advocacy groups. Opinions – often conflicting – about acceptability will affect both the demand and supply tracks of the conventional vision.

Acceptability of oil imports: the energy security issue

For the United States, and for developed countries in general, the question of energy security has moved from the agenda of defending national security to that of promoting foreign policy and domestic political interests.

In the 1970s and 1980s some governments of importing countries questioned the acceptability of increasing oil imports on the grounds of security of supply. Oil imports from the Middle East were expected to increase and the Middle East to become more unreliable and even threatening. The Iranian Revolution of 1978–9 was an example of a disruption of supplies by internal politics; the Iraq–Iran War and the Iraqi invasion of Kuwait were examples of disruptions caused by regional conflicts. The Arab oil embargoes of the end of 1973 were an example of an attempt to use oil exports as a tool of foreign policy. The price increases following the disruptions were interpreted as signs of a cartel to come. Increasing 'dependence' on imports was unacceptable to governments of importing countries, though few governments carried out significant policies to reverse the trend, except in France (through investment in nuclear power by the government electricity monopoly, Electricité de France), and in the United States (through the CAFE standards to improve automobile fuel efficiency).

Chapter 7 reviews the security question from the perspective of 2000. It seems more complex and more tractable. For all importers, the present market structure for crude oil of short-term, commodity prices offers more flexibility of response to temporary disruptions than existed in the 1970s. Markets will move oil from where it is relatively freely available to where supplies have been affected by disruption. For IEA members, the policy of stockpiling 90 days' worth of oil consumption and agreements to share supplies in emergencies is the first line of defence against shortages caused by temporary disruptions. The problem for the future is the shrinking share of the IEA membership in world oil trade and the exposure of Asian importing countries to the effect of disruptions on markets there. The possibility of an OPEC cartel sustaining oil prices above their competitive level in the long term seems remote, for the reasons given in Chapter 6. The experience of 1999–2000 suggests that there is a floor price below which all exporters will cooperate to pro-ration supplies, but the floor price is in the range of $10–15 per barrel. The real problem, as before, is the risk of a temporary disruption of supplies and the political damage which it might cause both as a result of the disruption of energy prices and in consequence of the international politics of the event.

Although the idea of the exporters uniting to use oil supplies as a sanction to secure political ends now seems remote from reality, some unusual combination of events can always surprise the world. In a narrowly balanced market, in which even Saudi Arabia has limited spare capacity, a major producer could achieve a temporary disruption of supplies and might achieve a limited political objective as a result. But the world of geopolitics has changed. Chapter 7 describes the different geopolitical environment for energy which has resulted from the collapse of Soviet influence, the US military impact in the Gulf War in 1991 and the military supply points and bilateral security treaties established for the continued US protection of oil supplies.

US protection is likely to continue. The United States faces permanently declining production of oil and gas and a permanently growing demand for clean fuels. Its nuclear capacity is being sustained by the relicensing of some old plants, but the majority are not due for relicensing in 2005–10. A difficult debate may then be expected on whether to relicense and embark on an expansion of nuclear capacity to meet the growing deficit of electricity generation and the mounting scale of the adjustment necessary to meet the Kyoto commitments or something like them. Meanwhile, or alternatively, the United States depends on the continual expansion of global oil and gas supplies to international markets to keep its energy costs low relative to other industrial countries. The US national interest will continue to demand a policy of protecting major oil exporters as far as possible against regional conflict and internal disruption. However, limiting that protection to the countries of the southern Gulf loads the main burden of external security of supplies for the United States on the US–Saudi relationship. Intemperate or accidental action on either side could disrupt that balance.

The same US predominance over the physical security of international supplies has reversed the risks that oil sanctions will be used by exporting governments as an instrument of foreign policy. Oil sanctions are now used by the United States and its allies against their own companies and citizens to prevent investment in, or trade with, the petroleum industries of certain oil-exporting countries: Iraq, Iran, Libya and Burma. The US administration has exerted its influence to promote oil and gas pipeline projects to enable independent Caspian exporting countries to avoid exporting through Russia. Its purpose here is not to serve energy objectives but to influence the foreign and domestic policies of these countries in directions desired by the United States. Other countries may in the future be targeted by the US administration or by powerful groups of US Senators or Congress members. These sanctions affect investment in the countries concerned and need to damage their economies to achieve their objectives. They therefore risk not only political reaction but also political instability.

For other oil-importing countries there is no feasible way to offer military protection to Middle East oil suppliers other than in cooperation with the United States. The EU and Japan are not involved institutionally or politically in negotiating the terms on which US security is offered or accepted. Nor are they major players in the Middle East peace process, which provides the other axis of US–Arab relationships. They can at best offer only symbolic alternatives to Middle East countries, which must define their relationships with developed countries positively or negatively around their bilateral relations with the United States.

Acceptability of carbon: the climate change issue

Chapter 8 puts the climate change issue in the context of a long history of interventions by governments to protect both human health and safety and the broader natural environment from damage. In many countries, governments' withdrawal from managing the supply side of the economy has been paralleled by their increasing regulation of the environmental impacts of satisfying the demand for goods and services of all kinds. Social control of the means of production is being replaced as an objective by social control of the manner of consumption.

There has been a steady growth of intergovernmental agreements to address environmental issues for which only international solutions will work: pollution of the seas, acid rain, ozone-depleting emissions and the transport of hazardous substances. These policies have developed against a background of growing public interest not only in matters affecting human health but also in the broad idea of sustainable development. What are new are the beginnings of international action to limit the emission of GHG, including the CO_2 inevitably emitted from the burning and combustion of carbon-based fuels – oil, gas and coal – and to protect the biological diversity of the planet as a whole.

For GHG, the UNFCCC of 1992 commits a group of governments of developed countries and 'economies in transition' (the 'Annex I' countries) to the aim of limiting their country's CO_2 emissions to 1990 levels by 2000 – an aim which obviously has not been realized in many countries. The Kyoto Protocol of 1997 assigns to the Annex I countries 'amounts' (the diplomatic equivalent of rations) of GHG, (including CO_2), for the period 2008–12. Individual governments in some major energy markets, such as Germany and the UK, have set more ambitious targets, building on reductions already achieved through reducing the use of coal for power generation. Some transnational companies, acting in anticipation of the policy trend, have set corporate targets for GHG reductions and are investing in the 'decarbonization' of their future energy supply business.

There are serious doubts about the implementation of the Kyoto Protocol. First, the economic growth of the countries concerned has diverged since 1990, and even since 1997. To reduce GHG emissions to the assigned amounts in the time period would require reductions from current 'business as usual' projections of GHG of about 30% in the case of the United States and 15% in the case of Europe.[7] Chapter 8 argues that this imbalance cannot simply be corrected by renegotiating another baseline; a long-term approach dependent on a single historic reference point will inevitably get out of step with a changing world. Second, even if implemented, the protocol would have little effect on long-term concentrations of GHG. Developing countries, which have

[7] These numbers are orders of magnitude. Different assumptions about non-CO_2 GHG, as well as about demand and fuel mix, would alter them by 2–3%.

made no quantified commitments, will before 2020 account for more than one-half of global GHG emissions from human activity. This disparity is a plausible (though not the only) reason why the US Senate has passed resolutions blocking ratification of the Kyoto Protocol unless developing countries accept a degree of commitment to GHG emissions reduction.

Proposals for trading allocated emissions between and within countries, and for a CDM to allow some GHG reduction commitments to be met by projects in developing countries, do not overcome these fundamental difficulties. Trading and CDM may reduce the cost of meeting the commitments in some cases, but would correct neither the imbalance of the commitments across Annex I countries nor the diminishing impact of their falling share of global emissions on the effectiveness of their efforts.

How any climate change mitigation policies affect oil demand relative to other fossil fuels depends on the design of national policies which in general do not yet exist. There are three competing and not necessarily consistent bases for such policies:

- Policies that were 'fair between fuels' – such as a new upstream carbon tax – might not affect the demand for oil much. More than one-half of oil consumption is in the transport market, where short- and medium-term alternatives either are not available or are very costly. How much of a burden the suppliers of oil would bear would depend on how oil prices, over which they have some control, react to the changes in such demand which would occur.[8] The users of oil in transport would bear a large share of the burden of adjustment costs (and loss of capital value) and would not necessarily benefit from the recycling of revenues (unless sums were earmarked for reducing the cost of alternative transport).
- Another basis of policy could be to protect or improve income distribution. Higher fuel costs may bear disproportionately on those

[8] See Mustafa Babiker, John M. Reilly and Henry D. Jacoby, 'The Kyoto Protocol and Developing Countries', *Energy Policy*, vol. 28 (2000), pp. 525–36; Ulritsch Bartsch and Benito Müller with Asbjørn Aaheim, *Fossil Fuels in a Changing Climate* (Oxford: Oxford University Press/Oxford Institute for Energy Studies, 2000), pp. 194–201.

for whom fuel is a large part of their essential expenditure and have little discretionary income. Policies aimed at changing the fuel consumption of the rich more than the poor would need to look at per capita allocations (rations) rather than equal taxation of carbon or energy content.
- Finally, policies aimed at allocating the GHG 'rations' to where they had most value would avoid attacking the transport sector in most countries. The prices (including taxes) users are prepared to pay for transport fuels in many countries indicate that carbon has a higher economic value there than in other sectors. Equity between transport users and other users would then be in question.

All policies based on taxes risk causing economic contraction unless the revenues are recycled, and even then there are winners and losers: energy-intensive industries and their customers pay the carbon taxes, labour-intensive service industries and their customers benefit if revenues are recycled by reductions of labour taxes. Redistribution is inevitably politically controversial. Faced with these difficulties, it is likely that many governments will, in this as in other environmental matters, make use of direct regulations on fuel use and fuel choice rather than rely entirely on so-called 'economic instruments'.

Until the design and severity of national policies for implementing GHG reduction commitments are clear, it is difficult to estimate the effects on the energy and oil sectors. The contradictions and difficulties of implementing the Kyoto system mean that clarification is unlikely to be achieved soon. But the process of policy-seeking will continue. The more the costs of adapting to climate change become evident, the more attention will be paid to GHG reduction policies. The longer effective policies are delayed, the more severe the policies are eventually likely to be to limit the threat arising from increasing levels of GHG concentrations.

The process of intergovernmental and internal negotiation on climate change will become a test of the idea of sustainable development. This idea presupposes that technology and well-designed policies can allow the achievement of both economic development and a much higher degree of protection of the environment – including the climate, and diversity in natural ecological systems – than would be likely under the 'conventional visions' extrapolated from the past. Testing these suppositions is the major challenge for both supporters and sceptics of sustainable development over the next 20 years. The willingness, and capacity, to start these social and industrial experiments resides in rich countries with dynamic economies which also have the means to promote the values of 'sustainability' at home and abroad.

The agendas for change which are likely to emerge may differ in North America, Europe and East Asia. They will be the result of a process of social and individual choices, policy designs and technology offerings from competing carriers of technology. These will include the oil and energy industries, but the choices are not limited to the supply side. Although fuel switching and new fuels can contribute to such solutions, the fundamental challenges are to the generation of demand. By 2020 climate change regulations, taxes, subsidies and public choices in developed countries are likely to define the demand to use fossil fuels as precisely as air and water quality and safety regulations define the supply of fuels today. Beyond 2010, this may attract attention again to the possibilities of nuclear power – possibly with different technical and management concepts.

Acceptable development

Concerns for sustainability go beyond the climate, and include resources, the diversity of species, and material and human capacity. 'Development' in this paradigm includes more than simple economic objectives. Life expectancy, health, education, the alleviation of extreme poverty and the easing of inequalities are part of the package. The World Bank has crystallized 'governance' as both an enabler for sustainable development and an objective in itself.[9] The success of market economies depends, according to this argument, on governments performing their core activities well, including the administration of justice. Respect for human rights, and the practice of democracy, are the best available mechanisms for ensuring 'good governance' in this paradigm. It can

[9] See World Bank, *World Development Report 1997* (Washington DC, 1997), 'Overview', pp. 1–13.

be summed up as seeing development as a process of expanding the real freedoms that people enjoy.[10] But the paradigm itself implies limits to the role of government. People are *directly* responsible for the success and nature of their civil societies, and of the emerging global civil society of which governments form a part.[11]

In the developed world, people with time and resources have an unparalleled opportunity to promote ideas that they share about the objectives of development and the means by which it should be achieved. Ideas can be spread through international agreements (like the UNFCCC) and intergovernmental institutions (like the World Bank) more easily because of the disappearance of communism from the international political agenda and the redirection of national bureaucracies away from microeconomic management. The global interlinking of cultures through education, entertainment, the media and the Internet is opening up a new way for internationalizing ideas. In Chapter 8 we argued that private sector companies operating in the developed economies, and accountable to investors in them, are also becoming important carriers of values, as well as of technology and finance. This is not a one-way, missionary endeavour from developed countries. Some values are universal. The new global connections and the new 'private' international relationships provide lifelines by which people anywhere can invoke international private support in their local struggles for political freedoms, better lives and cleaner environments. What they want, and what they create, will not necessarily be modelled closely on examples from America, Europe or Japan. The more they succeed, the more capacity they will have to make their own choices. Development will be freedom for them to be what they choose to be, not what is chosen for them in Washington, Tokyo or the multiple capitals of Europe.

---

[10] Amastya Sen, *Development as Freedom*, p. 36.

[11] 'For me, economic progress, scientific advance and public debate which occur in free societies *themselves* offered the means to overcome threats to individual and collective wellbeing. For the socialist, each new discovery revealed a "problem" for which the repression of human activity by the state was the only "solution" and state-planned production targets must always take precedence': Margaret Thatcher, commenting on her September 1988 speech on global warming in *The Downing Street Years* (London: HarperCollins, 1993), p. 641.

The road that enables all these ideas to move forward more rapidly than in the past is the information highway, where, thanks to the fax and the Internet, anyone can ride, whether governments like it or not. The freedom of the press has become a kind of freedom for everyone. Cheap and easy communication internally and across frontiers has already given NGOs an easy ride to the doors of governments. They have developed expertise and professionalism to take advantage of this, though different NGOs have different priorities and appeal to different constituencies. In the global market for ideas NGOs do not control all the traffic. The same road passes many other doors: investors and investment funds, customers and employees, citizens and local government as well as political parties and bureaucracies. Multinational companies need to deal with this traffic more directly than ever before. A whole dimension of international relations – the dimension of social and political values – has become privatized.

This approach to acceptable development silhouettes two possible roadblocks for the oil industry and those who deal with it.

The first problem is that of the governments of some oil-exporting countries where oil resources are being depleted more rapidly than they can be replaced by human capital or other forms of economic activity. Most of these are countries in the Middle East with large oil resources, where the process of consuming resources could continue for decades. Most of these countries have not been democracies. Their rulers and bureaucracies are supported by oil revenue and the patronage it finances, rather than by taxation and representative government. They have experienced some of the most rapid rates of population growth in the world in recent times. The need for revenue to sustain this system is one of the drivers of competition for oil markets that underpins the 'conventional vision' of increasing oil volumes (for some of the countries) with prices only slightly above the 1986–99 average. The idea of conserving oil resources for future generations could reappear in some countries, with an effect on the prices and investment cycles of oil and its competitors.

The second problem is located in the private sector. It is illustrated by the history of transnational companies that have achieved a high profile internationally for new oil or gas developments in Angola, the

Brazilian Amazon, Colombia, Indonesia, Burma, Nigeria and the Sudan. In many cases the problems have been addressed and improvements achieved. Their origins lay in one or more of the following circumstances:

- the projects involved serious environmental and social impacts;
- the host government was accused of extensive human rights abuses;
- democratic processes were absent or weak; and
- the revenues which oil revenues would bring to local governments were used for controversial purposes: to support corrupt elites, or to sustain civil war or foreign military adventures, and without attention to poverty and suffering in the country concerned.

Similar problems would affect transnational companies that take up, after the end of sanctions, the Iraqi government's offer of production-sharing agreements to expand its industry. Caspian and West African countries with new oil developments have the potential for challenges on some of these issues. Access to global communications ensures that these problems are widely known. Their potential effect on companies' performance attracts attention from international NGOs and 'ethical' investors and, because of them, the international investment community generally. To develop new supplies under these conditions may not be acceptable for the companies concerned unless they can satisfy potential international critics that their intervention is serving the wider developmental purposes of promoting the expansion of freedom for the people affected.

Strategies and risks

In this section the key points identified earlier in the chapter are connected to the interests of some of the main actors concerned. Those involved in oil have larger purposes: companies seek to produce value for shareholders who often are represented by investment institutions; governments seek to promote their national interest under the influence of public opinion; NGOs and excluded political constituencies promote a variety of ideals, using the global media and communications to persuade or threaten other actors. Some interests coalesce across borders and are the basis of intergovernmental agreements, the policies of transnational companies and the campaigns of transnational NGOs. For each group the 'conventional vision' presents different challenges and opportunities, the uncertainties underlying it present different risks, and the effect of others' choices change the strategic game. Simplifying to a few global scenarios may not help inform the decisions of anyone. Presenting all possible combinations would require unmanageably large numbers of cases. What follows is a very large-scale map of territory in which each enterprise or group has to plot its own course.

The transnational oil companies

The years 1998–9 saw a change in the structure of the large private sector oil companies.[12] Seven of the former top 15 companies (which account for nearly three-quarters of all private sector oil investment) were merged into three: Exxon, BP and TotalFinaElf. Shell, one of the largest transnational oil companies before and after the mergers, did not itself carry out merger activity but undertook a similar process of cost-cutting, strategic rebalancing and reorganization. The gap between the four largest companies and the rest widened. In contrast to the merger wave of the mid-1980s in the industry, many of the deals were based on share exchanges rather than cash raised by junk bonds. The logic was not (as in the 1980s) that share values did not represent 'real' values, but that new strategic visions, different portfolios of assets and strong management were needed for the changing circumstances of the private sector of the industry. The mergers were seen as driven by the need to cut costs, especially in downstream operations, to diversify upstream portfolios and to shift the balance of the four dominant companies towards future opportunities in gas.

The large companies' search in 1998–9 for a different future also reflects a change in their geographical interests. In the United States (but not in the world) the private sector oil companies were running out

[12] For an overview, see John V. Mitchell, 'Money, Management, Molecules', *The World Today*, vol. 55, no. 2 (Feb. 1999), pp. 18–19.

of oil and gas. In 1990 the top 15 private sector companies had produced 17% of the world's oil and the United States supplied 40% of their production. By 1997 these companies had held their share of world oil output, but the United States supplied only 28% of their production. For gas, the story was similar: the US contribution to these companies' gas output fell from 46% to 32% between 1990 and 1997. Between 1990 and 1997 the top 15 companies had increased their oil production by less, 1m b/d – an increase of 6% over seven years. A fall of over 1m b/d in their US oil production had offset half their increases in production of 2m b/d oil outside the United States. Their US gas production fell by 3bn cu ft/d while it grew by 9bn cu ft/d outside the United States. Among the 15, performance varied very widely, depending mainly on the weight of the declining US production in their portfolios. TotalFinaElf had and has insignificant US investments – a fact which gives it some freedom from US foreign policy.

The challenge which the 'conventional vision' presents to these companies is a menu of the following elements:

- *maintaining their share of world oil and gas production by expanding outside the United States* – very often into those countries where their operations may have impacts whose acceptability will be challenged in the arena of 'private international relations' or US foreign policy even if it is accepted by the 'host' governments;
- *preparing for a possible transformation of demand for transport fuels* as a result of competition among vehicle and transport technologies;
- uncertainty about the effect of *reform and restructuring electricity and gas industries* on the risks and returns of investment in long-term projects for gas expansion;
- uncertainty about the *medium-term investment policies of the major Middle Eastern oil-exporting governments*: how much expansion will those governments decide on and what role, if any, will they assign to foreign companies?

These challenges have to be met in a financial market where ambitious promises have been given to investors about sustainable rates of return and growth in value. In a world of inflation below 5%, 1–2% growth in volume and very little real price change, rates of return over 10% are unlikely to be achieved by all companies, just by continuous cost-cutting or clever asset deals. Technical and business innovations are inevitable: not all will be successful. The transnational oil companies cannot expect to escape the turbulent international competition for markets and technology and intersectoral competition for funds, which are transforming the electricity, telecommunications, automotive, aircraft, armaments and financial industries across the world.

Government agendas

The objectives of governments and the politicians who direct them in democracies are broad and various. The oil sector affects, and is exposed to, certain government decisions which are likely to be different in the future from in the past:

(1) Because of the recent *commitments to reduce GHG emissions*, OECD Annex I governments face the challenge of how to 'bend the trends' which are threatening to change the global climate. Will the Kyoto Protocol remain the centrepiece of policies in the medium term? When will a longer-term and more inclusive programme of international cooperation emerge? Will policies be implemented through demand management – punishing the consumers first – or through innovative alternatives to fossil fuels or consumption patterns which depend on their low-cost availability? The existing revenue-orientated tax structures on oil in Europe and Japan do not align well with GHG criteria. More government intervention will be needed to carry out GHG policies because the message of the 'conventional vision' is that developments in the energy sector are unlikely to provide fossil fuel prices so high as to induce the necessary changes automatically.

(2) *The shift in the balance of world oil and gas trade to Asia* gives new importance to the question of how to improve the capacity of Asian energy-importing countries to diversify their long-term supplies and build capacity to deal with short-term disruptions of oil supply. For example:

- bilateral or regional agreements to provide a legal framework for pipeline projects and operations;
- regional agreements for stockholding obligations and sharing supplies in the event of disruption; and
- regional agreements to prevent and manage oil spills and similar environmental catastrophes on shipping routes.[13]

The growing importance of Asia as a market, and of Asian companies and governments in relations with oil-exporting countries, may also affect the efforts of Europeans and Americans to promote their ideas of political, social and environmental acceptability of energy development and relationships with oil-exporting countries.

(3) *Oil-exporting countries' perception of their future* under oil prices similar to or below those of the past 20 years challenges them to improve the robustness of their economies and societies. Their populations are increasing faster than government revenue and there are few alternative sources of revenue or employment. This is primarily a question for the countries concerned. It is not limited to macroeconomics and is set in the context of the evolution of government and institutions in these countries, the policies of the United States and its relations with these countries, and the Middle East peace process.

(4) *New technology and limited geographical opportunities* are leading to the prospect of increasing private sector exploration and development activity in the Caspian, Central Asia and West Africa. Most of the countries concerned face difficult environmental, social and political challenges. The new oil and gas projects will have major impacts and may be challenged by international as well as local NGOs concerned with the environment, human rights and good governance. How to achieve 'acceptable' development in these countries is a common agenda item for the governments of the countries concerned, for the multilateral financial agencies, for the private sector companies responsible for high-impact operations, and for those who invest in the companies and their projects.

[13] There may also be a case for the coordination of efforts to prevent piracy.

NGOs and advocates of values

NGO interests are as varied as governments', and are less constrained by the constitutional processes which regulate government action and relations between states. Some NGOs enjoy recognition by the UN for the access this gives them to some level of intergovernmental meetings, and seek to achieve participation in the appraisal, if not decision-making, processes of multilateral agencies such as the World Bank. Such NGOs are likely to face questions about their own decision-making processes and sources of funds as they compete for legitimacy with the institutions of governments democratically elected in open societies. The burden of proving their right to invoke broad human values is rather different for NGOs in countries where governments are not elected, discussion and debate are limited by government, and human rights are not well protected by the rule of law.

Oil and gas use and supply trigger many environmental, social and humanitarian questions for NGOs. International NGOs' priorities are generally to minimize damage rather than to exploit the economic and social potential of oil or gas development and use. Although the economy of oil is changing, the inheritance of assets and habits is the starting point. In poor countries the search for the benefits of development carries an urgency for local people as well as their governments. Against the reference line of the 'conventional vision' and all the uncertainties and choices described earlier, the questions which businesses and policy-makers will increasingly debate with the NGOs are ones of prioritization and timing. What is unacceptable in the long term may be unavoidable in the short term.

There are three particular areas for debate between NGOs and their supporters and those who share their long-term objectives: in each the answers are likely to be specific to situations and transitory. Aspirations and tradeoffs will change if development brings rewards:

- *The mix of government regulation and private (including corporate) behaviour to be used to achieve their objectives.* It may be difficult to go far out of line with the balance prevailing locally across general economic and social activities: high regulation imposed on a society unaccustomed and unequipped to regulate efficiently and fairly may not produce good results. Relying on reputation and goodwill from those immune to public scrutiny and criticism may not work well either.

- *The scope of alternatives to be considered.* This is partly a question of timing: the long-term options for reducing GHG emissions are very wide, but in the medium term nuclear energy is seldom considered and never advocated (except by the industry itself) as an alternative to drastic attacks on demand.
- *The hierarchy of the next best.* Oil development in a particular country may have bad social and human rights results, but there is some scope for mitigating them through improving the practices of the transnational companies. If the effect of criticism is to keep transnational companies away from sensitive areas, the result may be that developments there are undertaken by operators less exposed to scrutiny and less sensitive to criticism. Alternatively, if the development does not take place, more supply will be drawn from the Middle East, where it is produced by governments who are protected against external pressures by their sovereignty, and many of which, for example, are not signatories to the UN conventions on human rights.

Conclusion

It is inevitably arbitrary to define a turning point or to draw a sharp line between new and old in affairs so complex as those of oil. The 'conventional vision' of continuing expansion of oil use with some loss of market share and without strongly escalating prices is a reasonable but fragile reference line against which to discuss the dynamics of the future. This book argues that these dynamics are sufficiently different that many images formed in the 1970s under different circumstances should be discarded or updated. So much has already changed. For example:

- Oil availability in the next 20 years (but not for ever) does not depend on resources but on investment, which competition among resource owners is likely to bring about, but not always at the right time.

- The onset of competition among technologies in the transport sector will change the long-term demand for fuels and services for transport.
- There will be stronger competition from natural gas for oil markets in power generation, in transport and in developing countries generally, but there are major questions of infrastructure investment and market structure to be resolved. Even if gas-to-gas competition becomes more common, gas and oil pricing will be inevitably intertwined where and when they compete.
- There is great uncertainty about future trends in gas prices, and in the distribution of value between users, distributors, transporters and producers.
- The energy security policy concerns of 30 years ago have been reversed. The development of global investment and supply gives cheaper security for importers than policies focused on reducing energy trade. It is the oil-exporting countries that are at risk through dependence on foreign oil markets. Sanctions have become a weapon of US foreign policy rather than a threat to oil importers. OPEC remains an organization within which most of the leading oil exporters try to limit competition among themselves in times of low prices, but lacks the potential to sustain a long-term cartel or to achieve a geopolitical identity.
- In the medium term, the Kyoto targets are unlikely to be met or even seriously attempted by many countries. However, over the next 20 years, policies to reduce GHG emissions are likely to deepen and broaden beyond those designed at Kyoto for the Annex I countries. The acceptability of increasing the use of oil as a fuel in the long term is already in question.
- In the next 10 years, high-impact projects for oil and gas developments will more and more be challenged because of their social and political, as well as environmental, effects. In the new, information-based and NGO-led arena of private international relations, companies will be held accountable for these effects.

Within this 'new economy', transnational petroleum companies are restructuring themselves, so far as they individually can or need, to accommodate a change in the balance of their operations from the United States (which *is* 'running out of oil and gas') to the rest of the world. Paradoxically, as the US government increases its dominance over the geopolitics of oil, the US companies are losing their dominance over the private sector oil business. For the oil industry, the future is one of accounting more to investors and to public opinion, and less to governments, for the acceptability of the developments on which future availability depends.

Index